Knowledge Machines

Infrastructures Series
edited by Geoffrey Bowker and Paul N. Edwards

Paul N. Edwards, *A Vast Machine: Computer Models, Climate Data, and the Politics of Global Warming*

Lawrence M. Busch, *Standards: Recipes for Reality*

Lisa Gitelman, ed., *"Raw Data" Is an Oxymoron*

Finn Brunton, *Spam: A Shadow History of the Internet*

Nil Disco and Eda Kranakis, eds., *Cosmopolitan Commons: Sharing Resources and Risks across Borders*

Casper Bruun Jensen and Brit Ross Winthereik, *Monitoring Movements in Development Aid: Recursive Partnerships and Infrastructures*

James Leach and Lee Wilson, eds., *Subversion, Conversion, Development: Cross-Cultural Knowledge Exchange and the Politics of Design*

Olga Kuchinskaya, *The Politics of Invisibility: Public Knowledge about Radiation Health Effects after Chernobyl*

Ashley Carse, *Beyond the Big Ditch: Politics, Ecology, and Infrastructure at the Panama Canal*

Alexander Klose, translated by Charles Marcrum II, *The Container Principle: How a Box Changes the Way We Think*

Eric T. Meyer and Ralph Schroeder, *Knowledge Machines: Digital Transformations of the Sciences and Humanities*

Knowledge Machines

Digital Transformations of the Sciences and Humanities

Eric T. Meyer and Ralph Schroeder

The MIT Press
Cambridge, Massachusetts
London, England

© 2015 Massachusetts Institute of Technology

All rights reserved. No part of this book may be reproduced in any form by any electronic or mechanical means (including photocopying, recording, or information storage and retrieval) without permission in writing from the publisher.

This book was set in ITC Stone Serif Std by Toppan Best-set Premedia Limited, Hong Kong.

Library of Congress Cataloging-in-Publication Data is available.
ISBN: 978-0-262-02874-5 (hardcover)
ISBN: 978-0-262-54785-7 (paperback)

Contents

Acknowledgments   vii
List of Abbreviations   ix
List of Illustrations   xi

1   A Digital Research Revolution?   1
2   Conceptualizing e-Research   23
3   The Rise of Digital Research   45
4   Aggregating People and Machines: Collaborative Computation   69
5   Distributed Data   95
6   Digital Research across the Disciplines: The Sciences and Social Sciences   125
7   Digital Research across the Disciplines: Humanities and Access to Knowledge   147
8   Open Science   175
9   Limits of e-Research   187
10   Knowledge Machines   197

Notes   225
References   237
Index   263

Acknowledgments

It is a great pleasure to thank the many people who contributed to this book—above all Bill Dutton, who led the two phases of the Oxford e-Social Science project and was invariably helpful and generous with his expertise. On the same project, we also enjoyed working with Jenny Fry, Matthijs den Besten, Marina Jirotka, Paul David, Annamaria Carusi, Rob Ackland, Arthur Thomas, and Grace Eden, among others. Sally Wyatt, Kathryn Eccles, Monica Bulger, and many others whom we cannot list here provided comment and criticism that helped shape the book. Christine Borgman's visits to the Oxford Internet Institute (OII) were the source of many insightful discussions, and Ann-Sofie Axelsson was a great help with Swedish e-research. At the OII, Emily Shipway, Tim Davies, Duncan Passey, Pauline Smith, and the rest of the staff provided a superb infrastructure. Our involvement with the doctoral students at the OII, including Christine Madsen and Lucy Power, provided us with perspectives on additional corners of e-research. The project was funded by the UK Economic and Social Research Council, JISC (United Kingdom), the Alfred P. Sloan Foundation (United States), the European Commission, the Research Information Network (United Kingdom), and Indiana University, among other institutions. Many interviewees have given time and provided expertise: these e-researchers have been excellent informants. We also thank the reviewers of the draft manuscript, whose suggestions were invariably constructive.

Many of the studies that went into this book were carried out collaboratively as part of various projects, and we have referenced them and acknowledged our many coauthors. This acknowledgment cannot do them justice, nor can it do justice to the greater detail of those studies, but readers interested in these more in-depth studies can find them via the reference list.

We also received useful feedback at many talks, particularly at the University of California, Los Angeles, Harvard University, and King's College London; in Amsterdam, Goettingen, and Canterbury New Zealand; and, of course, in presentations to our colleagues at the University of Oxford.

Finally, Ralph thanks his family, Jen, Sven, and Anja, who, among other things, were also great informants as e-researchers, and Eric thanks his wife, Michelle Osborne, for her unflagging support.

List of Abbreviations

AVROSS	Accelerating Transition to Virtual Research Organization in Social Science
BP	bipolar disorder
CENS	Center for Embedded Network Sensing
CERN	European Organization for Nuclear Research
CM	computerization movement
CSCS	Swiss National Supercomputing Centre
DISC	Database Infrastructure Committee
DSM	Diagnostic and Statistical Manual of Mental Disorders
EEBO	Early English Books Online
EEBO-TCP	Early English Books Online Text Creation Partnership
EGEE	Enabling Grids for E-sciencE
EGI	European Grid Infrastructure
e-SCL	e-Scholarly Communication Layer
ESRC	Economic and Social Research Council
GAIN	Genetic Association Information Network
GridPP	Grid Particle Physics
ISI	Institute for Scientific Information
IVOA	International Virtual Observatory Alliance
LHC	Large Hadron Collider
NCeSS	UK National Centre for e-Social Science
NGO	nongovernmental organization
NSF	US National Science Foundation
OBPO	Old Bailey Proceedings Online
OCI	Office of Cyberinfrastructure (US National Science Foundation)
PC	personal computer
RDC	Research Diagnostic Criteria
SAS	Statistical Analysis Software

SDSS	Sloan Digital Sky Survey
SIM	scientific/intellectual movement
SNA	social network analysis
SNP	single-nucleotide polymorphism
SPLASH	Structure of Populations, Levels of Abundance, and Status of Humpbacks
STIN	sociotechnical interaction networks
URI	Uniform Resource Identifier
URL	Uniform Resource Locator
VO	virtual observatories
VOSON	Virtual Observatory for the Study of Online Networks

List of Illustrations

Figures

Figure 2.1. Scientific styles.
Figure 3.1. Publications on collaborative computing topics, 1993–2012.
Figure 3.2. Number of news articles on big data.
Figure 3.3. e-Research publications and citations, 1993–2012.
Figure 3.4. Authors per publication.
Figure 3.5. Overlay map of e-research publications.
Figure 4.1. Complexity continuum.
Figure 6.1. Publication type by field, showing percentage of conference papers and journal articles only, drawn from the data set described in more detail in chapter 3.
Figure 6.2. Comparison of top-50 words in top-100 articles.
Figure 7.1. e-Research in the scholarly communication ecosystem.
Figure 10.1 (a–c). Styles and e-research.

Tables

Table 3.1. Publications Related to Early English Books Online
Table 3.2. e-Research Authorship by Continent, 2003–2012
Table 3.3. Countries and Citation Patterns of e-Research, 2003–2012
Table 3.4. Twenty-Five Prominent Institutions Publishing on e-Research
Table 5.1. Data Needed to Answer Key Questions in Psychiatric Genetics Case Study in Different Periods
Table 5.2. Features of Data in Six Case Studies
Table 6.1. Citation Rates by Field and Type of Publication

1   A Digital Research Revolution?

This book is about how the Internet has transformed knowledge. More specifically, it is about how digital tools and data, used collaboratively and in distributed mode—our definition of e-research, which we elaborate later in this chapter—have changed the way researchers and scholars in the sciences, social sciences, and the humanities do their work. We argue that there has been a transformation in how knowledge is produced, creating online "knowledge machines" that now underpin how research is advancing. In changing the world of knowledge, these knowledge machines are also changing the wider world of how research is accessed and used by a wider public and reshaping the physical and social world we live in.

A personal anecdote can provide a quick illustration. When one of us (Meyer) was an eleven-year-old student at a school in a small midwestern American town in the mid-1970s, his teacher had a contest throughout the year called "Question of the Day." Each day the teacher would post a new challenge to the class that would require extra effort to ferret out the answer to questions that were not usually to be found in the small school library. At the end of each school term, the student with the most points would get a prize. At the time, information was rare and difficult to find: finding the answer to "Whose portrait is on the $100 bill?" resulted in a trip to the local bank, requesting to see the bank's vice president, and asking him to retrieve a $100 bill from the vault. Only a few intrepid students were willing to put in the extra effort to find answers to the least obvious questions, whereas today any student in the classroom would have discovered in 10 seconds that the answer is Benjamin Franklin by accessing *Wikipedia* on their smartphones. Anything that is "Google-able" has essentially become instantly available to anyone.

The point of this story is not that "things were different when I was a kid!" but that the effort required to find information and data has changed dramatically in a single generation, and the emphasis is increasingly not on the skills needed to find information, but on the skills needed to assess, process, analyze, combine, and perform computation on data and information. The same applies, except even more so, to academic researchers. At least in the societies of the Global North, practically everyone—academics, students, and members of the public alike—has access to publications, documents, records, data, primary sources, and secondary interpretations of primary sources in a way that was not so before the emergence of the mass uses of the Internet in the 1990s. Whether this information is in the form of *Wikipedia* articles written by volunteer crowdsourced authors or the published results of medical trials made freely available on PubMed Central, the increase in accessibility is obvious to anyone with an Internet connection and a question looking for answers.

In this book, we try to understand not just how networked technologies have changed the ways we consume knowledge but also how new technologies have reconfigured the ways that knowledge is generated. Since the early days of computerization, the practice of research across disciplines and domains has been transformed by computational technologies, and, more recently, it has been transformed by the ubiquity of the Internet. At the most basic level, for example, search engines such as Google have become a dominant source of basic information not only among the general public but also for researchers. Regardless of fears that this so-called Google Effect will result in the "dumbing down" of research (Brabazon 2006; Waller 2009), the opposite is more likely since everyone connected to the Internet today has access to digital information in quantities unimaginable only a generation ago. As a research historian in one of our recent studies of humanities scholars put it, "For primary sources, I've now got more material than I will need probably for the rest of my lifetime" (anonymous informant, interviewed by project staff, September 2010, Oxford).

The possibilities of the Internet's advancing research go far beyond basic information search, however. From massive physics collaboratories such as the European Organization for Nuclear Research (CERN) Large Hadron Collider (LHC), which relies on high-speed networks to distribute data around the world for analysis, to web-based resources such as Galaxy Zoo,[1] which is harnessing the power of citizen science to classify galaxies from the Sloan Digital Sky Survey (Lintott, Schawinski, Slosar, et al. 2008; Raddick, Bracey, Gay, et al. 2010), the Internet is increasingly embedded in the practices of doing science. Likewise, from genealogy enthusiasts researching family history (Yakel 2004; Crowe 2008) to individuals researching health information (Eysenbach, Powell, Kuss, et al. 2002; Hesse, Nelson, Kreps, et al. 2005), both researchers and members of the public can use a variety of digital technologies and access a vast range of resources via the Internet in the interests of research. These technologies are knowledge machines in the sense that they are instruments for manipulating data and thus for advancing and making use of research. They provide a common focus for researchers at the research front, spurring them to improve the capability of these tools to perform operations on data. Because the instruments are, moreover, part of a single online system that is growing ever more powerful and integral to the research process, we argue that even if it may be too early to say that these machines are responsible for a "revolution," we can say that they are responsible for the most important transformations of the sciences and humanities in our time and for the foreseeable future. This conclusion makes it worthwhile to examine the nature and implications of these transformations.

A Typology of Internet Research

In this book, our main focus is the production of new knowledge, primarily as it is undertaken by university researchers. But it is useful to put this topic in a broader perspective of what "research" can mean in the context of the Internet—if only to set the stage for defining e-research, which, we argue, is a subset of research. Indeed, as we shall see, one issue raised by the digital transformations of research is whether these broader and narrower topics have become more difficult to disentangle. We must in any event provide an overview of the broader landscape and can distinguish five forms of Internet research as follows (based on Meyer and Schroeder 2013).

The first form of Internet research is *about the Internet as a* **social phenomenon**. This approach, common at institutions such as the Oxford Internet Institute and among members of the Association of Internet Researchers, can be thought of as the social science *of* the Internet, answering questions about the Internet's role in changing how people form and maintain relationships, communicate with governments, learn, play, and interact. Examples of this approach are legion and are well represented in a number of recent handbooks on the topic (e.g., Hunsinger, Klastrup, and Allen 2010; Consalvo and Ess 2011; Dutton 2013).

The second form of Internet research is about how the general public, students, and researchers *use the Internet as an* **information source**. This is what people are doing when they consult *Wikipedia*, search for health information, find research papers, undertake student projects, and engage in other information-seeking behaviors. An extensive literature on information seeking, some of which goes back decades, has more recently engaged with information seeking on the Internet. It is particularly well developed in the library and information science field and is frequently a topic in the field's leading journals such as the *Journal of the American Society of Information Science & Technology* and the *Journal of Documentation*, among others.

The third form of Internet research is about *the Internet as research tool and* **research method**. In this case, the Internet can be seen as a means for measuring and discovering patterns of human behavior, online or offline. Online survey research, webometrics, virtual ethnography, and myriad other new, adapted, and emergent methods enabled by the Internet fall into this category. Several recent handbooks provide guides to many of these online research methods (Fielding, Lee, and Blank 2008; Hesse-Biber 2011; J. Hughes 2012). Some of what we discuss in this book falls partly within this category.

A fourth form of Internet research is about engineering and computer science: the research into advancing the **underlying technologies** *that allow the Internet to function*. Much of this work is happening in industry and in industrial/academic partnerships and involves the design of both the hardware (including routers, switches, and cables) and the software to keep the Internet running, growing, and improving. In Europe, the Future Internet Assembly[2] is just one example of efforts to advance research into this area, as are a number of sections of the professional organizations IEEE (Institute of Electrical and Electronics Engineers) and ACM (Association of Computing Machinery).

Defining e-Research

Our focus in this book is the fifth form of Internet research, which is somewhat less well known than the first four and involves understanding *the Internet as an underlying* **research technology or infrastructure** *that is* enabling advances in research practices across multiple disciplines and domains. The label sometimes attached to this perspective is *e-research* (Borgman 2007; Dutton and Jeffreys 2010b), which we define as **the use of digital tools and data for the distributed and collaborative production of knowledge** (Meyer and Schroeder 2009b) and which can be seen as a subset of digital research that takes place primarily at various research frontiers. Most of the research in this area is not about Internet users and uses or about the Internet as object, technology challenges, or a new medium to enact research methods. e-Research is rather about creating new knowledge, including scientific, social scientific, and humanities knowledge, which we distinguish in due course. Much of this e-research, which encompasses the more specific categories "e-science," "e–social science," "digital social research," "digital humanities," and "cyberinfrastructure," is not *about* the Internet at all. Instead, as we shall see, e-research in this sense has become inextricably bound up with advances in Internet-related capabilities that have led to the availability of a globally connected network of machines and people (or more precisely, *researchers*), with wide-ranging consequences.

Although all five forms of Internet research are at least somewhat overlapping and dependent on each other, e-research specifically, in advancing knowledge production and thus advancing the research front across many disciplines, has the potential for some of the most lasting implications of the Internet as it has developed over the past several decades. It is important to note that throughout this book we take a broad view of the Internet, which according to this view encompasses not only the network but also the devices and tools connected to it. A purist might argue that "the Internet" refers only to the network of networks that uses the various layers in the Transmission Control Protocol/Internet Protocol (TCP/IP) model to transport data. Our broader view may be less technically accurate, but it is more accurate in terms of how to think about the Internet's sociotechnical network. This idea of the Internet includes not only the underlying network devices and connections but also all the devices, sensors, applications, databases, networked tools, data, and content residing on these technologies, for without them the network itself is uninteresting and not powerful at all.

In this book, our main concern is to understand how knowledge generation changes with the application of networked research technologies to academic research. Nonacademic research also relies heavily on the Internet as research infrastructure, including research in industry, nonprofit organizations, and governmental research institutes. This type of research, however, is largely beyond the scope of this book, though it should be noted that there is vast potential for social research based on the data collected by commercial and public bodies. Mike Savage and Roger Burrows (2007, 2009), for example, have argued that these so far largely untapped sources of transactional data, much of which is collected via the Internet, is one of the greatest potential sources of sociological research data but also threatens to make traditional research methods based on sampling and in-depth interviewing obsolete (for a critical view of Savage and Burrows's argument, see Schroeder 2014). We touch on the potential of these so-called big data that emerge from commercial organizations at several points in the book.

In our discussion, we use the terms *research* and *knowledge* interchangeably, but *knowledge* refers only the products of research (i.e., not simply the everyday sense of knowledge as "I know how to get there"). And again, we also use the term *research* (or *knowledge*) to encompass science, social science, and humanities, but we separate science from nonscience in accordance with Schroeder (2007b). The demarcation of science from nonscience and labeling the study of the social as scientific are of course fraught with difficulties, and we return to this topic on a number of occasions.

We describe the major changes occurring in the world of research through the application of digital technologies, focusing on how the Internet and other networked technologies have enabled collaboration at a distance and the kinds of tools and data that have enabled new research practices. The main questions we want to address are, *What difference have digital technologies made to research, and how have they changed the direction and the practice of research?* There is a difficulty with these questions, and it is just as well to flag this problem at the outset: it is increasingly difficult to identify a single overall transformation brought about by e-research or networked digital technologies separately from the general effects of the Internet. Nevertheless, we argue that it is possible to pinpoint some consequences of a more specific set of research technologies (e-research) on knowledge production, with wider implications for the role of science and knowledge in society.

With the benefit of hindsight, it can be seen that the boundaries of e-research were initially marked by several highly visible—at least within academia—funding initiatives in the United Kingdom, United States, European Union, Australia, and elsewhere. As we have already indicated, the core of e-research can be defined more precisely, as opposed to the vagaries of various short- to medium-term funding programs that we discuss in later chapters. Over time, some of those funded programs have become part of normal practices in academic research, and many researchers are likewise increasingly engaged in research that is clearly relying on shared and distributed digital tools and data for research even when they are unaware of developments in the various e-research funding programs or even of the term *e-research*. These "accidental e-researchers," as we refer to them, represent a growing number of researchers across the disciplines who are using shared and distributed digital tools and data for their research—not because they have been attracted by targeted funding initiatives, but because their research questions and scientific objectives require computational approaches. We see this relationship in several examples provided in this book, and in others we see that new technologies have colonized the research front in the disciplines in question and that their use has become commonplace. Throughout the book, we give illustrations of how specific transformations have taken place—in how research is organized, how research is shifting relationships among academic disciplines, how researchers' practices are changing, and how computational approaches contribute to producing new knowledge. Although it is impossible to capture a process that is still ongoing in a comprehensive way, we argue that it is nevertheless possible to provide insights into the implications of this process to date and thus also to provide signposts for the future.

Approaches to Understanding e-Research

To provide further context for the topic of how digital technologies are transforming research, a brief overview to the approaches to this topic and some examples of the kinds of transformations that fall within its purview are in order.[3] Despite some notable exceptions (e.g., Borgman 2007; Dutton and Jeffreys 2010b), this transformation has received little attention, in part due to the fragmentation of disciplinary perspectives that address this phenomenon. The digitization of scholarship has been analyzed from a variety of (inter)disciplinary perspectives. These perspectives include the sociology of science and technology studies (STS), as in studies of technological infrastructures (Jackson, Edwards, Bowker, et al. 2007); media studies and how scholarly communication changes the publishing process (Boczkowski and Lievrouw 2008); political-institutional perspectives on the social shaping of technology (Dutton 2011); information and web science, as in studies of sharing data (Borgman 2007); the fields of economics of innovation and research policy, where various issues surrounding intellectual property rights in multi-institutional collaborations have been discussed (David and Spence 2003); and computer science and computer-supported collaborative work, where the difficulties of distributed work have received considerable attention (Olson, Zimmerman, and Bos 2008). Apart from the different topics covered, these disciplinary perspectives often have different aims: for example, contributing to the design of systems or addressing policy issues surrounding e-research or providing a critical approach to the claims about its transformative nature (Schroeder and Fry 2007; Fry and Schroeder 2010).

This is only a partial list of topics covered from these (inter)disciplinary perspectives and their approaches. Many others might be listed, and we make use of them throughout the book. However, a key implication is that the diversity of perspectives, which is a characteristic of Internet studies as a whole, is likely to persist. One way to avoid a fragmented picture of digital transformations of research is to focus on a single substantive question and then to illuminate this question from various disciplinary angles. In the case of the digital transformation of research, we argue, as already mentioned, that this question must be, How is scholarly knowledge changing as it moves into the online realm? Put differently and in line with our definition of e-research, the question is, (How) Do the distributed and collaborative uses of digital tools and data open up new directions for research?

We address this central question from a variety of perspectives. One point to anticipate is that the digital transformation of knowledge would ideally be contextualized in a larger picture of how information and knowledge are being transformed by the Internet and related new media in society at large. These broader ideas about social change, often associated with terms such as *information society* and *knowledge society*, are beyond the scope of the book, but we come back to them in the concluding chapter in relation to the somewhat more specific role of scientific or scholarly knowledge in society.

In order to preview our argument, we offer here some brief examples of a few common shifts in how research is changing at the level of knowledge on the one hand and in scholarly practices on the other (we discuss these examples and others in greater detail in later chapters). Once we have done this, we will be able to make a start on developing our theoretical argument in chapter 2 about how scholarly knowledge is being transformed with e-research and how a synthetic approach can overcome the limitations of various disciplinary approaches to the topic.

Transforming Research

There are literally thousands of examples of e-research projects of varying size and impact. Some of these projects have arisen in response to the various funding programs sponsored by research councils (see Dutton and Jeffreys 2010a); some are part of the attempts to build the e-infrastructures or cyberinfrastructures supporting science and research (see Barjak, Eccles, Meyer, et al. 2013); and yet others represent "bottom-up" or "accidental e-research" (see Meyer and Dutton 2009), or e-research that has arisen independently when individuals or groups build their own tools for wider uptake or sharing and collaboration. The examples we give here illustrate the range of e-research, how the technology is used in research, and the changes in research practices involved. They also offer a foretaste of the more detailed studies that make up the bulk of the book's chapters and prepare the ground for the next section, where we preview our argument about how e-research is transforming the sciences and the humanities.

Physics and Large-Scale Data

We can begin with uses of the Grid and distributed computing in physics, which provide a good illustration of how the scaling up of organizational complexity and the scaling up of technological complexity go together. The LHC at CERN is a good example, and here we can focus on the UK e-research part of this effort, which is called Grid Particle Physics (GridPP) —a collaboration of particle physicists and computer scientists at nineteen UK universities and Rutherford Appleton Laboratory. The UK part of this effort is connected to a series of larger European-led and global collaborations to enable a Grid computing effort by means of large-scale research funding programs (here, the European Commission–funded Enabling Grids for E-sciencE [EGEE][4] and European Grid Infrastructure [EGI][5]). The LHC is a large research instrument—a piece of technology or engineering or a machine—that, put simply and crudely, smashes particles together. The experiment, using a recent example, consists of postulating a smaller particle than those that have been identified previously (the Higgs Boson)— this is the theoretical part. The practical part of the experiment is designed to undertake observation and measurement of these postulated particles. Computing plays a major role in this effort because the data generated by the experiment are large scale, requiring high-throughput processing. The Grid is used to handle this volume of data using a multitiered structure so that the parcels are delivered into what are called Tier 1 machines, which further parcel them out to Tier 2 machines, and so on.

The organization of these complex and distributed computing operations is highly bureaucratic in the hierarchical arrangement of this parceling out. However, it is also nonbureaucratic insofar as rather than having a top-down command structure, the assignment of tasks throughout the hierarchy takes place on the basis of memoranda of understanding that are arrived at relatively democratically (Shrum, Genuth, and Chompalov 2007). The reason this way of doing research deserves the label *e-research*— apart from this organization of sharing—is that the computing operations are scaled to highly parallelize the task of analyzing the data.

This effort also has its national constituent parts. The United Kingdom's GridPP,[6] for example, uses a sizeable share of the data from the current LHC experiment and participates on all the tiers. To this end, the UK physics community has had to develop an organization to dedicate computing power to analyze these data in coordination with other physicists from around the world who are involved in exploiting the LHC data (see Zheng, Venters, and Cornford 2011). The key feature of the LHC and GridPP to highlight is that physicists have for some time been a community that has had to coordinate research on a large scale and in teams with a high division of labor but also a high degree of mutual dependence and task certainty (Whitley 2000). Physicists are thus well placed to undertake complex tasks that parallelize computational data analysis across multiple levels and in a distributed manner. Hence, GridPP illustrates how physicists have had to adapt their organizational practices to fit around the highly distributed way in which their data need to be analyzed with—in this case "grid"[7]—computing tools. In any event, particle physics illustrates that highly complex organizational and technological infrastructures are needed to cope with data on a scale that is unparalleled in other areas of research.

Public Engagement via Citizen Science
A different example of computation on a large scale, though in this case with much less organizational and technological complexity—a much more "flat" and simple structure—are volunteer computing efforts, such as Galaxy Zoo. The Galaxy Zoo project makes use of citizen scientists to classify galaxies from the Sloan Digital Sky Survey (Lintott, Schawinski, Slosar, et al. 2008; Raddick, Bracey, Gay, et al. 2010) via a public website.[8] Unlike GridPP, which is using distributed computation to analyze data, Galaxy Zoo relies on distributed human cognition to classify data, a task for which human brains are still more suited than computer algorithms. Citizen scientists are shown an image of a galaxy and then are given a set of fairly simple choices: Is the galaxy smooth, or does it have features such as a disk? Is the galaxy clumpy or not? Does it appear to be a disk viewed from its edge? Choosing among these options isn't a very difficult task for the human brain, but it is very difficult to perform computationally. In addition, human classifiers don't have to be highly skilled because each image is shown to multiple classifiers—if most of them agree in their classification, the likelihood that their classification is correct is quite high.

The scientists involved in the project report that not only has this project surpassed their expectations but is making real contributions to science (Chris Lintott, interviewed by project staff, July 19, 2011, Oxford). Papers have been published with citizen scientists as coauthors,[9] and the project has resulted in new discoveries, including the now famous Hanny's Voorwerp, a previously unknown astronomical object discovered by Galaxy Zoo participant and Dutch school teacher Hanny van Arkel.[10] What we see in this case is how a relatively simple technology—a website—can enable the distribution of tasks to a large number of volunteers rather than to computers, as in the GridPP example. However, organizing the task in such a way that it attracts many volunteers and organizing the collection and analysis of the data are by no means minor tasks. Nevertheless, once the tool was created for one application (astronomy), it has required less of an effort to develop a platform called Zooniverse that can be extended to distribute other tasks that require thousands of volunteers. In short, this kind of volunteer effort enables labor-intensive tasks to be distributed via the Web, but in an organizationally and technologically "flat" way (as opposed to the tiered, hierarchical way in the particle physics case) in that the effort consists of thousands of volunteers logging into a website via their personal computers (PCs) and their responses being aggregated within a single database. Both the particle physics case and the Galaxy Zoo case, moreover, illustrate another argument to be made later: that technologies for research—in other words, knowledge machines—that are developed for one purpose can easily be adapted for another.

The Complexities of Data Sharing

A different way of harnessing digital resources shows that it is nevertheless not a straightforward task to bring distributed data together. The Center for Embedded Network Sensing (CENS) was set up to see how the availability of small and cheap sensing devices could be used to monitor environmental conditions. These sensors were deployed in the field in various settings and could provide streams of data back to scientists operating in the lab.[11] One of the findings of this project (Borgman, Wallis, Mayernik, et al. 2007) is that participants from different domains viewed the data generated by the sensors differently. For instance, the engineers on sensor projects were interested in performance data from the sensors (e.g., packet transmission information, battery state, fault detection); scientists were interested in scientific data (e.g., water temperature, wind speed, and acidity or basicity values); and data managers and analysts were interested in contextual data that could influence the scientific data (e.g., motor speed of the boat pulling the sensor, calibration data, environmental conditions of the equipment connected to the sensors).

Unlike physics and volunteer computing for astronomy, then, this kind of distributed computing effort demonstrates a different kind of complexity: building data libraries that not only accommodate the needs of various kinds of researchers but also allow for future use and reuse of the data. This complexity, as Christine Borgman and her colleagues (2007) have argued, requires an understanding of the entire data life cycle and an ability to integrate data from multiple sources. In this case, we can see that the main effort is not in the organization of the computational task (as in GridPP) or in the coordination of the task among many volunteers (as with Galaxy Zoo), but rather in the implementation of an extensive set of tools that require coordinating digital data over a heterogeneous set of instruments so that they provide a long-term data resource that is as consistent as possible. The major effort in this case lies in shaping organizations and their data practices.

Data practices have moved into the forefront of the challenges in e-research (Borgman 2015) for more reasons than simply the heterogeneity of data among different types of researchers. A different type of challenge lies simply in sharing data on a large scale among different groups where this larger scale is a precondition for making scientific advances. A further useful illustration here is the Genetic Association Information Network (GAIN) project, which was a US effort starting in 2006 to do large-scale genome-wide association studies of originally six selected data sets related to a variety of medical conditions (Manolio, Rodriguez, Brooks, et al. 2007). One of this project's innovative aspects was that unlike in previous research, where investigators would have exclusive use of their data for a period of time (often one year from the end of data collection), in GAIN the "results [were] made immediately available for research use by any interested and qualified investigator or organization" (Manolio, Rodriguez, Brooks, et al. 2007, 1048), including organizations such as pharmaceutical companies. The attraction to researchers was the promise of large-scale, high-quality genotyping of their research subjects in exchange for their contributions of DNA from blood and diagnostic information.

The risk for researchers in participating in GAIN was that they could be beaten to publication on results from their own data because other researchers could analyze those same data at the same time. In other words, the contributing investigators had just a six-month exclusive publication window, after which anyone using the data could publish (Meyer 2009). This model of high-speed, high-stakes collaborative science suggests the potential to greatly speed up scientific discovery, but at a risk to individual scientific achievement. Nevertheless, the incentives whereby funding was made conditional on sharing data in a common pool in this case demonstrated that large-scale collaboration to produce data sets across research teams is possible. However, the result has been that genome-wide association studies science requires even larger data-sharing efforts to make progress in finding links between genes and medical conditions.

The GAIN project illustrates large-scale data sharing for a specific research goal. But e-research also consists of much more ambitious efforts to develop data infrastructures that link different types of data at a population-wide level as a long-term infrastructure for research. Where the effort involves sensitive data, as with certain types of medical or social research, e-research poses the additional challenge of requiring the trust of populations that these data will not be misused in any way. One example of such an effort is the Swedish e-research program, which has embarked on precisely such an undertaking. One of the main preconditions that favors e-research in Sweden is the availability of many high-quality data sets about the Swedish population that have been maintained for a long time and have recently become the responsibility of the Swedish National Data Service.[12] There are at least two further reasons why Sweden is uniquely well equipped to embark on such an effort. First, each person living in Sweden has a unique personal identifier (the so-called person number) that is frequently and commonly used in everyday life and for record keeping by the government and other bodies (for example, health records but also validation of a credit card). It allows researchers to make links between different types of data, which is typically difficult in other countries due to the heterogeneity of different types of data sets. Second, there exists a high level of trust between the Swedish population and researchers and government due in part to long-standing laws concerning (especially computer-supported) data protection and in part to a number of widely publicized debates around the use of data in research.

Researchers are currently revisiting these factors in the development of e-research capabilities at the Swedish National Data Service and more generally among Swedish researchers, who are well aware both of the unique circumstances in Sweden with respect to the quality and potential of its digital data as well as of the need to maintain the trust that has been established and that will have to be maintained if this potential is to be fulfilled (Axelsson and Schroeder 2009). What we see illustrated in this example is that when sensitive data are involved, the issue is not just organizing large-scale data collection and sharing (as with GAIN, where the data are anonymized) but also how collecting data about whole populations requires a high level of trust between researchers, public authorities, and the public when these data derive value from linking different data sets and where this linking poses potential risks for research subjects. These conditions do not obtain in many countries, and it is an open question whether they will continue to do so in Sweden. For the moment, however, the relatively high level of trust in Sweden and the availability of many long-term data sets provide better preconditions for this type of e-research than in most other countries. Put differently, the key in this case is not merely large-scale organizational efforts (as for GAIN), but also longer-term social enabling conditions. We return to these examples in more detail in later chapters.

The Challenges of Infrastructure across the Academia–Industry Divide
Sharing data and linking large-scale, heterogeneous, and sensitive data sets are not just a challenge in academic research. In recent decades, particularly in the life sciences, there has been increasing collaboration at the interface between academia and the private sector, which raises further issues for research spanning different institutions. Although such collaboration between industry and academia has not been extensively explored in the literature on e-research, it is bound to become more significant, particularly where large-scale public funding is involved. Here the SwissBioGrid project provides a good example inasmuch as it consisted of a collaboration from 2004 to 2007 between academic institutions and the private sector—specifically pharmaceutical companies (den Besten, Thomas, and Schroeder 2009)—that was successful in the form of a demonstrator project, albeit at a level of modest complexity and funding.[13] The project involved a number of academic partners, including the Swiss Institute of Bioinformatics at the University of Basel and the Swiss National Supercomputing Center. The collaboration consisted of two tasks: proteomics data analysis and "virtual screening" to find new drugs for dengue fever. The core goal of these tasks was to develop tools for harnessing distributed-computing power (in part by using many idle PCs overnight) for such aims in life sciences research, which succeeded in demonstrating the technical capabilities required. In developing the software, SwissBioGrid drew on the CERN physics collaboration for some of the "middleware"—that is, the tools for distributed-data management.

On the one hand, the project demonstrated that successful e-research collaboration between the private sector and academic institutions is possible and that distributed computing in the life sciences can provide useful research tools. On the other hand, despite its success, the project was limited insofar as there was at the time no larger infrastructure such as a national e-infrastructure that could be used to sustain the tools and data over the longer term. Such Switzerland-wide efforts at resource sharing have since emerged in the shape of the Swiss National Grid Association. Yet even if the project could not be sustained within a larger infrastructure, it demonstrated the value of harnessing many PCs to perform a computationally intensive task, as with GridPP. Unlike GridPP, however, the main effort in the SwissBioGrid project was not in coordinating the computing power of distributed organizations, but rather in coordinating several partner institutions to join and in tailoring the software to the task of performing the analysis. Here, though, we can see the limitations of e-research in the case of projects that take place without being embedded in longer-term technical and organizational structures, as in the Swedish e-research effort, which began with the creation of a national infrastructure for research.

Changing Everyday Practices and Supporting Researchers

e-Research requires not just new tools and organizational structures but also changes in researchers' everyday practices. An example here is the Structure of Populations, Levels of Abundance, and Status of Humpbacks (SPLASH) project (for other examples of efforts that required changing practices, see Schroeder and Spencer 2009). This project was "one of the largest international collaborative studies of any whale population ever conducted ... [involving] over 50 research groups and more than 400 researchers in 10 countries" (Calambokidis 2010, 7). Using photo-identification techniques that exploit the fingerprint-like nature of humpback whale flukes (tails) to identify individual animals, scientists were able to start to answer population-level questions about the whales that individual scientists working alone could never hope to answer (Meyer 2009). Through this large collaboration, scientists were able to construct a much more complete picture of the current populations of Pacific humpbacks and to understand more about their migrations and seasonal movements.

The project is a good illustration of how a large effort is required in changing research practices—in this case, making the transition from non-digital to digital images. This effort is not so much a question of technological complexity, but rather of how many researchers can make the transition to handling and classifying digital images. Further, even though the amount of data was not large by the standards of research in astronomy or in physics, it nevertheless presented a daunting challenge to federate and analyze these data, particularly given the large-scale and collaborative nature of the project (Calambokidis, Falcone, Quinn, et al. 2008; Meyer 2009; Calambokidis 2010). And again, making this change across a number of organizations, as with the CENS example, required a major level of coordination.

Changing practices requires not only making the transitions in technology but also expanding the effort to maintain support of these practices.

The latter demands not just technological and organizational resources but above all—in some cases—expertise dedicated to the task. A good illustration is the Virtual Observatory for the Study of Online Networks (VOSON), a web-based tool to study online social networks.[14] This tool was developed at the Australian National University beginning in 2005 to promote "webometric" approaches that measure, among other things, the visibility of websites in relation to the links (or hyperlinks) between them (Ackland, O'Neil, Standish, et al. 2006; Ackland 2009). This tool allows social scientists to identify patterns of online links between, say, environmental movements or between political parties in order to understand their offline networks and activities, among other things. VOSON has been developed as an open-source tool with a Creative Commons license that allows researchers not only to use it but also to build upon and extend it. Thus, VOSON has been used by a group of researchers worldwide who themselves belong to a wider community of researchers employing social network analysis (SNA)—a community that includes researchers from beyond the social sciences and that often uses computationally intensive methods.

VOSON continues to be developed and has been used in a number of publications, and it can be seen as part of a burgeoning engagement in e–social science or digital social research. From the researchers' point of view, apart from requiring skills in the analysis of websites, the tool requires only that they create an account and follow the instructions that are provided about how to use it. Nevertheless, the challenge with tools such as VOSON is to maintain them (for example, in making the tool compatible with other tools) and to provide support to clients/users, who often have queries. Unlike Galaxy Zoo, which requires minimal user support, web tools such as VOSON require engagement with and ongoing support for a user community, a task that requires time as a key resource in addition to software expertise.

Engaging Communities in the Humanities

In yet other e-research efforts, such specialist expertise may not be required, and the goal is to engage a community of contributors. We have already seen this in the example of Galaxy Zoo, which has been successful in enabling amateur contributions to astronomy. But volunteer efforts are not restricted to science. In the humanities, too, a number of projects have harnessed the power of "crowdsourcing." One such project is *Pynchon Wiki*, a wiki created with the purpose of annotating the works of the contemporary American novelist Thomas Pynchon (Schroeder and den Besten 2008).[15] One element of Pynchon's work is that his writing is full of allusions and obscure references, which invites a detective-like effort at producing annotated editions to help the reader understand these references.

Whereas Pynchon's novels published before 2006 had been annotated in book form over a period of many years, the members of the *Pynchon Wiki* effort were able to annotate the entirety of his new novel, *Against the Day* (Pynchon 2006), in mere months after its release. The creation of this wiki involved hundreds of contributors. Apart from speeding up the process, *Pynchon Wiki* has reconfigured how this effort takes place: contributors find which parts of the novel have already been annotated, fill in the gaps that remain, and carry out additional discussions if the entry for a particular term has already been "finalized." Thus, it is not so much that literary studies becomes transformed as a discipline via this wiki, but that the wiki allows for amateurs or nonacademics to contribute rather than just individual academics.

Scholarly practices are further transformed insofar as the wiki allows some multimedia annotation (images, sound clips) to be added to text, but, more importantly, the task can be scaled to a large and open community of contributors, which can find out where annotation is needed and coordinate with others to fulfill that need. Like Galaxy Zoo, then, this tool allows a task to be highly distributed. Also like Galaxy Zoo, this web tool has subsequently been extended to a number of wikis for other Pynchon novels and wikis for the novelist David Foster Wallace and the Beatles. Such web-based research technologies depend on enabling the scaling up of human effort much more than organizational and technological effort, and they can be applied to a variety of labor-intensive tasks.

A final example here illustrates that e-research is appropriate not just for knowledge-generating tasks on a large scale or for large communities of scholars or amateur contributors. The community of scholars that specializes in the study of Tibetan and Himalayan texts is quite small (500–700 scholars worldwide) and comes from several disciplines, including religious studies, anthropology, and philosophy. A number of projects have been developed to digitize the texts and images that were previously difficult to access due to the distributed nature of the collections, some of which are in remote locations (all material for this case is based on Madsen 2010). A few individuals have devoted themselves to reorganizing this community's work, and the challenge is to embed these projects within an organizational form that can sustain them (perhaps with an institutional base and continued funding and staffing).

This small multidisciplinary specialism has not been dramatically transformed by the digitization of resources, although key staff members responsible for digitizing texts have had to become experts in using the required software. Moreover, although the main scholarly work in this area remains the interpretation of texts (and thus the main change in scholarly practices is in access), some scholars have also used the digital versions of texts to undertake computational linguistic analyses, such as frequency of terms, which was not possible before now. Hence, this is one way in which the research front has changed. Another is that in the case of a relatively small community of specialist scholars, unlike the large numbers that have participated in Galaxy Zoo and *Pynchon Wiki*, accessing remote text requires the dedicated work of a few individuals with the requisite computing skills and, as with VOSON, a high level of effort to develop and maintain the tools.

These brief examples illustrate the range of ways research technologies are enabling the reconfiguration of research practices and the emergence of new research questions across the disciplines. We return to most of these examples in more detail in due course, but for now they suffice to set the stage for the rest of our discussion, with the exception of one important topic.

Research Visibility and Access

The changes discussed so far have been on the side of producing research or new knowledge. What about the consumer side, or how knowledge is received, which is typically discussed within information science or the subfield of science communication? One change taking place here (see Meyer and Schroeder 2009b) is how these two sides are becoming linked: academics are becoming increasingly aware of how their publications and other outputs are being received, in general because academic outputs are increasingly online and thus subject to measurement. Downloads can be counted, webometrics can be performed on the number of links to a particular web page, and scholars are aware of how their work is accessed by means of search engines such as Google and Google Scholar. The automation of these ranking and measurement mechanisms produces a feedback loop whereby the "visibility" of scholarship feeds into academics' behavior (scholars make sure they are working on topics that are highly "visible"), which ensures that their work is more frequently accessed (see, for example, Willinsky 2005a). Thus, a new system is added to existing systems of research assessment or evaluation, attaching online visibility to traditional scientometric rankings of individuals and research institutions (by number of citations or comparative rankings of universities, departments, journals, and so forth).

On the consumer side, these changes also entail new practices among researchers in how they access materials. Students, for example, search for information using search engines and increasingly rely on *Wikipedia* articles. Although scholars use specialized searches in journal and other databases, they also similarly use general-purpose search engines such as Google to find material on specific topics and may be led to articles on the open Web (Rieger 2009). Again, note the feedback loop: the more *Wikipedia* is relied upon for, among other things, students' and academics' research, the higher it will appear in search results. In this way, the boundary between academic research and the public is becoming somewhat blurred (Meyer and Schroeder 2009b). This brings our argument full circle, for the proof that science and scholarship is being transformed cannot simply be in the creation of knowledge but must also be evident in its impact and wider role in society. And although it would go too far to say that e-research is contributing to a "knowledge" or "information society," the various research fronts (e-research fronts in particular) are certainly becoming more visible in the online realm, which is, of course, expanding not just among researchers, but in everyday life and in society at large.

To say that the transformations of digitizing research are merely changes in scale or speed is therefore misleading. When does quantitative change turn into qualitative change? Science and research more generally advance and gain more power over the natural and social environments (Schroeder 2007b). Perhaps beyond this advance, it is possible to gauge paradigm shifts or scientific revolutions, as Thomas Kuhn and Karl Popper would have it, only with the benefit of hindsight. Nevertheless, it is possible to chart how change takes place in a variety of ways—reorganizing disciplinary boundaries, changing day-to-day practices, shifting funding priorities, making more extensive and intensive use of computation, and more. These effects fan out across research fields, and gauging the cumulative gain in the power of research will tell us when a digital transformation has taken place. This is difficult to do across the board for all disciplines, though we have given a wide variety of examples in this first chapter and will give others in the chapters that follow. Instead, it is possible to examine advances in particular areas of research and in particular research communities by reference to the state of the art and then to draw these analyses together to show how knowledge is being reshaped more generally.

Overview of the Book

In chapter 2, we discuss various ways one can use existing theories to conceptualize computational approaches to research, in particular developing the idea (following Ian Hacking) that a few "styles of science" are self-authenticating and that e-research computational approaches support and intensify these styles. Further, it is possible to understand how these computational approaches are promoted by conceptualizing researchers as part of sociotechnical interaction networks. Combining these two arguments, we can say that research technologies are a strong driver for advances in research, but so, too, are computerization movements for embedding computational approaches to research across the disciplines, and thus say that technological advance and sociotechnical interaction networks reinforce each other. These ideas constitute the theoretical core of the book, which we pursue at the hand of various cases and broader aspects of research.

In chapter 3, we map the growing prominence of e-research and related approaches to generating knowledge using data from funders and from publication databases. The following two chapters present several case studies from our own research: chapter 4 focuses on the ways that machines and people are aggregated via the tools of e-research, and chapter 5 looks at the possibilities and challenges in sharing research data. Chapters 6 and 7 take on the issue of disciplines and how disciplinary differences influence the modes and manners of engaging with computation and algorithms to advance research. Chapter 8 turns to the question of openness in science and how digital tools create new possibilities and challenges for open science. Chapter 9 examines some of the bumps on the road toward e-research that we have observed in our research. Finally, chapter 10 revisits the book's central questions and ties together the material from the previous chapters.

The data and cases discussed in this book are the result of a series of research projects funded over the period 2005–2012 (now projects about "big data"), which used a variety of methods to gather data about computational approaches in research across the disciplines. Major funders include the UK Economic and Social Research Council (ESRC),[16] JISC (formerly Joint Information Systems Committee, United Kingdom),[17] the Alfred P. Sloan Foundation (United States),[18] the European Commission,[19] the Research Information Network (United Kingdom),[20] and Indiana University.[21]

We collected data using a variety of methods, both qualitative and quantitative. The qualitative data include individual and focus-group interviews with hundreds of participants,[22] spread across the research projects noted in the previous paragraph, that center on various aspects of how computing is enabling changes in research practices and in the nature of research across the disciplines. Interviewees ranged from domain scientists in the physical sciences, social sciences, and arts and humanities to computer scientists as well as officers and other staff at research-funding agencies. We also used observation and participant observation data from time spent with researchers in the field.

Quantitative data supporting the conclusions in this book include survey data, scientometric data from Scopus and ISI, and webometric data gathered from the public Web. We draw these quantitative and qualitative findings together and aim to present a more complete picture of the transformations of the research landscape than has been possible in these individual studies and projects. The sources of data and methods are noted throughout the book in relation to particular case studies as appropriate, though more details can also be found in the publications cited.

2 Conceptualizing e-Research

Before we can look at specific cases of e-research, we first need to establish a framework for understanding how these research technologies reshape knowledge production. The examples given in chapter 1 provide a backdrop for this conceptualization or theory. The theory presented in this chapter will subsequently inform more detailed discussion of these and other examples in the chapters to come. In presenting this theory, we proceed from a general view of science, which stresses the internal workings or the autonomy of science via research technologies, to a more external view, which concentrates on social forces "on the ground." Both an internal perspective and an external perspective on e-research, we argue, are needed: it must be shown how scientific knowledge advances, but because e-research obviously requires expensive and organizationally complex sociotechnical systems to be put into place, it must also be shown how the social momentum is generated to implement and sustain these systems.

Styles of Science

First, we can note (as already mentioned in chapter 1) that one feature shared among e-research efforts is that they consist of online research technologies with a digital component, though what this component consists of varies among particular e-research projects (such as volumes of data or enhanced processing power). These technologies support the production of scientific knowledge, which Ian Hacking (1992, 2009, 2012) argues consists of six "styles" that are "self-authenticating" or "self-vindicating" and that make for the autonomous nature of science—or for why science needs no further "foundations."

Hacking's argument draws on the work of A. C. Crombie (1994), who provided a historical account of these styles, which are (1) mathematical or by postulation; (2) experimental exploration or the measurement of

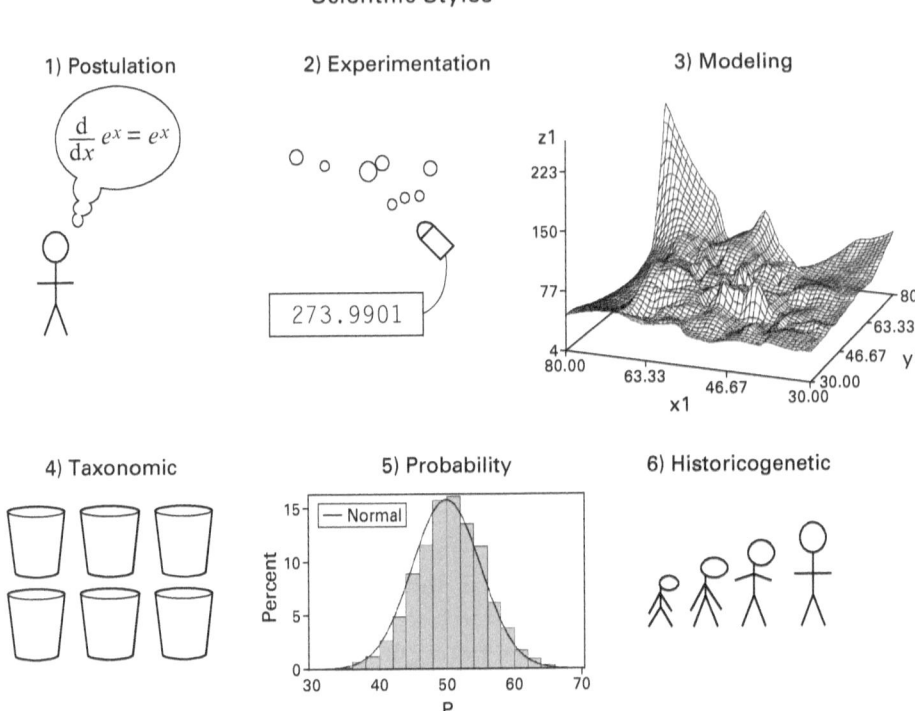

Figure 2.1
Scientific styles.
Source: Based on Hacking (1992).

complex observable relations; (3) hypothetical modeling or the construction of analogical models; (4) taxonomy or the ordering of variety by comparison; (5) probability or statistical analysis; and (6) historicogenetic explanation (Hacking 2002, 181–182, 2012) (see figure 2.1). These six core styles rarely exist in their purest form in modern science because many scientific pursuits combine multiple styles (Kwa 2011). For instance, modern genetics and genomics rely not just on the core principles of how genetic developments have been derived historically (as with some of Darwin's ideas) but also on discovery of the functions of genes and their expression using statistics, taxonomy, and possibly experimentation (in the case of nonhuman genetics, for instance, in experimentation on rats and mice) as well as modeling.

It can easily be seen how these six scientific styles apply to e-research or to how knowledge is produced using digital technologies. Thus, we can categorize e-research efforts by the digital tools and objects—the digital components—they use. So, for example, some e-research engages in *in silico* experiments (thus, style 2, experimentation), whereas other e-research models the heart or climate change (style 3, modeling). Still other e-research, such as wiki-style efforts, build taxonomies (style 4) of biological processes and the relations between them. Statistical or probabilistic analysis (style 5) underlies much of the work done on data sets, including genetics research that looks for statistical significance in genetic samples. The first and sixth styles are less evidently connected to e-research because theoretical postulation has no obvious links to computational processing, but we will see in a moment that mathematics (or algorithms), which are the essence of postulation, are in fact central to e-research. The historical derivation of genetic development (again, part of Darwin's theory of natural selection) is exemplified by efforts such as tracing genetic development over long periods of time.

Hacking's idea of "styles" is not without critics (Kusch 2010), which he has ably addressed (Hacking 2012). Nevertheless, one of the strengths of his idea is that scientific styles cut across the boundaries of disciplines, and so we will also see that the e-research technologies that support these styles are unevenly distributed across disciplines and fields of scientific endeavor. The main criticism that Hacking has failed to address is that his account is not very realistic in terms of showing how *modern* science is distinctive and why it has transformed the natural and social environment more powerfully; in short, his ideas about science are ahistorical and asociological.[1] One of us has given such a comparative historical and sociological account (Schroeder 2007b), however, drawing on the work of Randall Collins and others, which focuses both on mathematics and on research technologies as well as on the institutions of science and technology.

Collins argues that mathematics is a necessary condition for the rise of modern science: "the distinctiveness of the network of mathematical practitioners is that they focus their attention on the pure, content-less form of human communicative operations: on the gestures of marking items as equivalent and of ordering them in series, and on the higher-order operations which reflexively investigate the combinations of such operations" (1998, 873). In contemporary terms, this activity sounds much like the algorithms performed by computers (even though Collins is thinking about the role of mathematics in science during the past four hundred years or so). We can add to Collins's ideas about "mathematical rapid-discovery science" his second condition for the rise of modern science: research technologies, or "the lineage of techniques for manipulating formal symbols representing classes of communicative operations" (1998, 874). The advantage of invoking the role of mathematics in this way is that it avoids seeing mathematics in abstract terms as somehow transhistorical or transsocial. Mathematics is simply a "repertoire of reliable methods of transforming one set of gestures into another" (Collins 1998, 874), a way in which researchers communicate novel results to each other in a demonstrable way, especially if the results can be endlessly reproduced in a tangible way by research technologies—in this case, networked computers.

Before we go on to elaborate the notion of research technologies further, we want to point out that Collins's ideas tie Hacking's styles with research technologies: Hacking's "postulation" or "mathematical style" can be equated with Collins's ideas about the centrality of mathematics, but this style can also be a seen as the contentless operations that computers undertake (algorithms, the mathematical or logical instructions for computers), which, on a large scale and with large volumes of digital data, are at the core of e-research. Here we can anticipate later chapters by noting parenthetically that there is some truth in the assertion of the importance of Cloud computing and big data (although these are also buzzwords—we measure this "buzz" in chapter 3): Cloud computing increases the amount of data that can be stored and processing power that can be accessed (via shared networks) to manipulate these data, and big data provide (via shared networks) large sources of digital data from social media or databases created from digital interactions. e-Research uses networked computing (digital tools) to manipulate digital data by means of calculations—algorithms to produce knowledge. Why is this process so important in advancing research? To answer this question, it is necessary to elaborate the role of research technologies.

Research technologies, as mentioned, led to the rise of modern science. Collins argues that research instruments enabled "high-consensus rapid-discovery science" (1998, 532–538) in Europe from about the year 1600 onward. This argument stands the conventional wisdom—that progress in science leads to more powerful technologies—on its head. Collins argues instead that novel technological artifacts drive the advance of scientific knowledge. It is easy to think of examples: new and improved telescopes, microscopes, galvanometers, and today, computers that lead to scientific discoveries because they allow more powerful representations and manipulations of the physical world.

The reason why research technologies make for "high-consensus rapid-discovery science" is that they make it possible for scientists to use equipment to manipulate phenomena, improve the research equipment to do this, and move on to new phenomena: "What was discovered was a method of discovery; confidence was soon built up that techniques could be modified and recombined endlessly, with new discoveries guaranteed continually along the way. And the research technologies gave a strong sense of the objectivity of the phenomena, since they were physically demonstrable. The practical activity of perfecting each technique consisted in modifying it until it would reliably repeat the phenomena at will" (Collins 1994, 163). What is new in modern science is "secure knowledge" and "a train of new results" (Collins 1994, 157). As we shall see, this process can readily be applied to the use of digital technologies and data in research.

Collins thus avoids an idealist account of science and technology: the advance of scientific knowledge here takes place only in relation to the physical and social worlds conjointly, never just in the realm of ideas. This view meshes well with Hacking's philosophy of science, whereby science is the "adventure of the interlocking of representing and intervening" (1983, 146), a pragmatist and realist account of the relation between scientific knowledge and the physical or natural worlds. One of us has developed Hacking's ideas by arguing that technology is "the adventure of the interlocking of refining and manipulating" by means of physical instruments or tools (Schroeder 2007b, 9).

In Collins's terms, science is driven by technologies, laboratory apparatus in particular. Collins describes a process whereby there is an "outward flow of lab technology," and this technology is "exported into the lay world" (1993, 315): "One machine gives rise to another in a genealogy of succession: by modifying the past machine, or by cloning it from another in the same laboratory, or by a kind of sexual reproduction recombining parts from several existing pieces of equipment" (1994, 164). In this way, research technologies ultimately gain legitimacy by means of their diffusion into consumers' everyday practices (Schroeder 2007b). These arguments about how technology (and science) transform everyday life are beyond the scope of this book: here we are dealing only with how the world of knowledge and research are being transformed. Knowledge, however, ultimately reshapes everyday life beyond research, a connection that also applies to e-research.

To underscore this point, we can take a brief excursus to ask, Does this argument apply to research fields that are less dependent on research instruments, such as most of the fields in the social sciences? Collins argues that it does, but to a far lesser extent because the social sciences have been slow to adopt the use of research technologies. Here we can think of the use of recording technologies in sociology, which has become commonplace only relatively recently, in part because of the expense, quality, and bulkiness of recording equipment (Lee 2004). The same can be said of the uses of computers for large population data sets: since the US Census Bureau adopted commercial computing in 1951 (Ceruzzi 1999), the scale and scope of the entrance of these data into broad circulation and the ability to manipulate them more powerfully have undergone a sea change, much of it in recent decades.

Nevertheless, as Collins notes, even if these research technologies have led to advances in certain social science specialisms, they have hardly led to consensus or to rapid discoveries. Indeed, the social sciences are notorious for their lack of consensus and their factionalism (Collins 1994)—or, to use Richard Whitley's terminology (which we turn to shortly), for their generally low degree of mutual dependence and high task uncertainty. Thus, whether research technologies will lead to greater cumulation of knowledge in the social sciences remains to be seen. In any event, once a research technology is in place, it is typically refined for improved performance, is extended to new domains, and takes the form of combining an organization with a network of technologies. Or, as Thomas Hughes (1994) argues, technological systems acquire a "momentum" of their own over time.

A second impact of research technologies, which Terry Shinn and Bernward Joerges (2002, see also Shinn 2008) have noticed, is that they advance knowledge by transferring knowledge between different disciplines or domains. We have already encountered such movement between fields briefly in the examples of Grid computing (access to remote high-performance computing resources) and tools for crowdsourcing tasks in chapter 1. e-Research tools, especially software, like other research technologies thus often act in the manner of "multi-level, multi-domain intelligibility devices" (Shinn and Joerges 2002, 244). Whether this is happening *in practice* with regard to the technologies of e-research is a question we discuss more fully later in this chapter.[2]

Research technologies, according to Shinn and Joerges, reconfigure disciplines and extend the reach of knowledge by diffusing across boundaries, a "practice-based universality" (2002, 245). Put differently, these instruments disembed knowledge practices and provide translations between different knowledge domains. These changing practices are evidenced in different ways throughout the cases in this book, but there are also common elements, such as the shift toward putting data online and the increasing manipulability of these data by means of software tools and algorithms.

Not all e-research fits the definition of research technology, and not all e-research exemplifies the changes described here. Within the sociology of science and technology, however, there is a continuing debate about the relation between "local" knowledge and the universality of science, and we argue that the research described here is evidence of the translocality of knowledge even if this potential universality is still in the process of being instantiated. In any event, technology, ironically enough, has been largely missing from social science accounts of how e-science (or indeed any science) is reconfiguring knowledge. Shinn and Joerges, like Collins, argue that research technologies have been critical to the advance of scientific knowledge, though they locate the advent of their main social significance in the late nineteenth century, when universities and firms began to have research laboratories.

Further, there are extensive discussions in science policy about how to organize scientific efforts that are increasingly complex in scale and scope and that entail collaboration that is both interdisciplinary and distributed among different institutions and geographical locations. e-Research, we argue, is a useful place to examine these changes on the research frontier. Shinn and Joerges assert that research technologies are "generic" or "open-ended general-purpose devices" (2002, 212), instruments that can be used across a range of disciplines. Such technologies are initially developed outside of established disciplines and institutions, and in this sense they are "interstitial" and "disembedded," removed from particular groups' interests. Thus, they are able to create a new language and way of representing phenomena that transcends particular disciplines and institutions.

At the same time, these all-purpose devices then become "re-embedded" in multiple local contexts, spreading this shared language and means of representing phenomena across them. These "practices are independently repeated and are multiplied in innumerable environments. This is not the objectivity born of pure reason or the *experimentum crucis*. Objectivization is instead built up through collective practice which is structured around effect-producing materials and procedures. ... Objectivization is cumulative and practical" (Shinn and Joerges 2002, 244). Or, again, research technologies consist of "concrete ... practices" rather than of abstract scientific knowledge or cultural representations of technology. These practices include "design, hands-on-construction, endless tinkering and analysis to probe the deep principles of devices, adaptation to improve performance, explorations and controls to determine the extent to which a generic device can be generalized, trials and modifications to check whether the processes of generalization hold, and transferring apparatus into a local niche environment for tailoring and operation by end-users" (Shinn and Joerges 2002,

217). Recall here the "interlocking adventure of refining and manipulating" included in the definition of technology given earlier and Collins's ideas about how research instruments are modified and reproduce results.

In this way, as illustrated by certain tools in the past, "specialty groups" in different disciplines "learned to communicate and came to see aspects of their problem domains in the light of how ... [the] instrument represented and dealt with the physical world" (Shinn and Joerges 2002, 217); once these innovations had taken hold within the specialty groups, the terminologies and practices would have then "moved outside the research–technology nexus and into many professions and countries" (2002, 218). These ideas apply Collins's outward flow of laboratory technologies to the uses and spread of research practices. In other words, they allow us to see how the process of generating knowledge is being increasingly driven by knowledge machines.[3]

e-Research has been promoted largely by technology developers, whose aim is to develop tools that can be applied to a range of disciplines and purposes. These generic devices allow manipulating data and other digitized material across networks. Among the characteristics that these technologies share are a means of finding and classifying relevant data or other resources with the help of digital identifiers (including using ontologies, tagging, and putting data into formats that allow search and retrieval); a means of providing secure access to the data and tools (through mechanisms such as job submission processes and portals for the coordination and control of shared digitized resources); standardized software protocols; and ways of distributing the manipulation, storage, and communication of data and other digital research resources between different computers (from PCs to high-performance computers) via networks.

Shinn and Joerges think that the "meta-methodologies and meta-artifacts belonging to generic instruments, and which are re-embedded in local, narrow-niche devices, operate like passports" (2002, 244). e-Research technologies fit this definition insofar as they represent "a form of instrument design that consciously takes into account maximizing the variety and number of end-users whose local technologies can incorporate key features of a research-technology template" (2002, 212–213). "Communication between institutionally and cognitively differentiated groups of end-users," Shinn and Joerges say, "eventually develops" (2002, 244). Whether the technologies in e-research do this in practice is, of course, an open question, not least within the e-research community itself. And we can anticipate that there will be both reluctant users and practices that spread across communities at various research fronts.

How users and practices vary can be pinpointed by noting that different disciplines or fields have different objects that they focus on at the research front. e-Research, like all research, consists of a constantly moving front whereby scientists orient themselves toward the leading edge of the community to which they are contributing and where they compete to advance within this community beyond the current state of the art in relation to a particular object or phenomenon (Gläser 2006). The distinctiveness of e-research in this regard is that the means by which researchers orient themselves is that it creates a physical (technological) networked system that is aimed at this community's state of the art, which therefore provides a persistent focus for all researchers working in this area. Whether this effort is successful or not cannot be determined in advance, but the e-research component has the advantage of having enhanced visibility and of being able to be built upon—in addition to making use of more computationally intensive tools and accessible manipulable data. (One way to gauge whether e-research constitutes an advance in this respect is to ask whether the opposite holds: whether stand-alone computers—or, indeed, lone researchers—are equally or more powerfully able to achieve such advances, in which case e-research would not be considered an advance.)

Putting these two characteristics together (the core of a networked research technology or instrument and a community oriented to a common research front), e-research involves the creation of a coherent network oriented toward the state of the art of the community investigating a particular research object, a network that can be built upon and that provides computationally (mathematically) powerful manipulation of data about this object within one of the styles of science. (One proviso is that in parts of social sciences and in the humanities, these research technologies may not represent and intervene, as in the natural sciences, but rather generate new questions—for example, at the leading edge of the humanities.)

There is yet another way to describe this research front and a group of researchers' focus on a common core: in e-research, there is a physical core consisting of a network (the Internet and the Web) and manipulable data that are maintained and used by a community or by groups of researchers. It is possible to distinguish this physical core from the organization of people that supports and makes use of it (although these social forces extending and making this core more organizationally robust are also needed to advance the research front, as we shall see). In each case, this core produces new computationally intensive ways of doing research, whether by sorting materials, processing them, or performing numerical analysis (which are among the styles of science). It has been argued that these physical cores—research instruments—typically enable scientists to mobilize around them (Fuchs 2001a, 306, 330), an argument that applies to e-research, which consists of the development and greater use of computational tools and methods. Even if these tools and methods are initially used only in specialized niches, it is likely that they will strengthen over time because—unlike the (nontechnological) organization of people—they can be added to, extended, applied to other domains, and reconfigured. We document this strengthening and the resulting advances in certain research fronts but also more broadly. We observe how this strengthening occurs around research instruments and how these advances change the contours of the research landscape, but this is also a moving target because e-research is still science—and knowledge—in the making. Thus, we can map these digital transformations of research—and how networked tools create new technoscientific organizations (that always have a physical component) and promote the intensification of computational ways of advancing the research front around knowledge machines.

This process does not always involve research technologies moving from the natural sciences to the social sciences and beyond; they can also move from the social sciences to the physical sciences. One example of technology that has made this reverse journey is SNA, in particular measures of centrality (Freeman 2008). Linton Freeman argues that although sociologists using SNA techniques had been working in this area for decades, only in the past 10 years have "huge amounts of readily accessible relational data and ... easy access to large-scale computing power" (2008, 10) made it feasible to scale measures of centrality up to the levels required for the techniques to move into physics and biology, where they have subsequently grown rapidly.

This explanation of how scientific knowledge advances in a variety of domains and how research technologies move rapidly from one area to the next applies to modern science generally. What about digital research? When are digital materials available for this kind of mathematical, computational or symbolic manipulation? At this point, it becomes necessary to distinguish between different fields or disciplines and to notice that digital data and how they are manipulated by digital tools vary significantly.

Richard Whitley's (2000) ideas about degrees of mutual dependence and task certainty in different disciplines can be applied here. His central argument is that researchers in certain disciplines are more mutually dependent on the actions of their peers for scientific progress, whether that means agreeing to standards (such has been done in the astronomy community), collaboratively building large research facilities (such as the LHC in physics), collecting longitudinal survey data sets in the social sciences (such as the American National Election Surveys,[4] which date back to 1948, or the British Household Panel Survey,[5] which has been done since 1991, or the World Internet Project,[6] which has been measuring Internet penetration and use since 1999), or digitizing and sharing humanities collections (such as Europeana[7]). Whitley distinguishes between functional dependence, whereby researchers have to use their peers' results or procedures, and strategic dependence, wherein researchers have to convince their peers of the value of their research. It is easy to recognize that some of the examples just given, such as physics, fall into the category of functional dependence, whereas others, such as the humanities, fall into the category of strategic dependence.

A second dimension that Whitley identifies is task uncertainty, which can be technical insofar as techniques can be more or less well understood and reliable or strategic in the sense that problem formulations and significance can be more or less stable. Task uncertainty (or, conversely, task certainty) refers to the extent to which the steps of research and the tools to accomplish those steps are agreed and clearly understood or if the very aims of research are open-ended or subject to different interpretations.

One of the questions we must ask ourselves in the area of e-research is, To what extent do different levels of mutual dependence and task uncertainty in research-oriented fields and disciplines lend themselves to engaging with the tools of e-research and with computation via different organizational forms? Is it the case that because standards are more easily implemented in fields with high mutual dependence and low task uncertainty that top-down approaches to building large infrastructures are more successful in these fields than in others? Likewise, are fields with low mutual dependence and high task uncertainty more likely to engage in bottom-up activities, building many small, nonstandard, highly specialized, task-specific tools?

We return to these questions at several points throughout the book. For now, however, we can note that Hacking's styles and Whitley's model of mutual dependence and task certainty fit together in the sense that they help us understand different aspects of scholarly research: Hacking helps us to see that the ways of knowing the world are autonomous and limited to a number of "styles," and Whitley helps us to understand how the socioscientific characteristics of disciplines and the objects of their research can likewise be simplified for analytical purposes to a few core types of social organization. These two perspectives operate at a fairly high level and leave out many details and overlaps in how science and research operate in practice.

The relative power of technology and social forces to drive change in science and research is an ongoing debate in the study of science and knowledge. But they are also clearly complementary: research instruments or technologies have driven the advance of science and shaped society (even if in this book we address mainly changes in knowledge and research and address the larger implications for society only briefly in the conclusion). From the invention of early technologies such as microscopes and telescopes to the more recent construction of large research technologies such as particle accelerators, space telescopes, and large-scale genotyping facilities, the availability of these scientific instruments has no doubt influenced the types of scientific questions that are pursued, and we provide examples of this influence. However, it is also evident in e-research that resources must be mobilized, a task that includes not just funding but also the mobilization of expertise and organizations in order to develop and sustain these research technologies. These sociotechnical interaction networks or scientific/intellectual movements also shape the directions of knowledge production.

The question of how research technologies drive knowledge or shape the scientific questions that can be pursued is important, first, because, as discussed earlier, science is cumulative and builds on previous knowledge. As Isaac Newton famously wrote to Richard Hooke, "If I have seen further it is by standing on ye sholders of Giants" (Newton 1959, 416). Science relies on prior knowledge that has become durable and accepted as the foundation of new research. One implication of an increasingly computational approach to research, as we shall see, is an increasing scientization of disciplines that have historically been less cumulative, such as much of the work in the social sciences and humanities. Research technologies give momentum to this shift toward scientization in part as a result of the application of mathematical approaches to previously unquantified and unquantifiable topics.

Social Informatics

A mesolevel way to understand how technologies work in the world is to use a social informatics perspective (Kling 1999; Kling, Rosenbaum, and Sawyer 2005). This perspective moderates a technology-centric, scientific realist, and technological determinist perspective (Schroeder 2007b), although not going as far as to embrace the social constructivist position of actor–network theory (see Latour and Woolgar 1979; Latour 1987) or the social construction of technology (see Pinch and Bijker 1987; Bijker 1995, 2001). Social informatics emphasizes that "[information and communication technologies] do not exist in social or technological isolation" (Lamb, Sawyer, and Kling 2000, 1614) and thus allows us to understand sociotechnical configurations as being driven by a combination of social and technological factors. Indeed, one of the contributions that we make here is to go beyond these perspectives and thus advance the sociology of science and technology, especially when we return to these debates in the conclusion.

We define social informatics as a field of study that tries to understand sociotechnical configurations in the world by focusing on the relationship between the social, or "socio," and the "technical," without a priori privileging either side (what one of the authors has elsewhere called focusing on the hyphen in the "socio-technical" [Meyer 2014]). This analytical neutrality is one of the ways in which social informatics differs from technology-centric perspectives (such as those implicitly expressed in many computer science and engineering studies) and from strong social constructionist perspectives (such as those explicitly adopted by many in the sociology of science and technology). From a social informatics perspective, in any given sociotechnical configuration, technological and social/organizational factors may be equally important, but frequently one or the other is more dominant in shaping the configuration at different points in the history of a sociotechnical system.

One of the useful tools to emerge from social informatics is the sociotechnical interaction network (STIN) approach, following Rob Kling, Geoff McKim, and Adam King (2003), which is "an emerging conceptual framework for identifying, organizing, and comparatively analyzing patterns of social interaction, system development, and the configuration of components that constitute an information system" (Scacchi 2005, 2). The STIN framework is a more fully developed version of what Kling earlier referred to as web models (Kling and Scacchi 1982; Kling 1992b): "Web models conceive of a computer system as an ensemble of equipment, applications and techniques with identifiable information processing capabilities ... an alternative to 'engineering models,' which focused on the equipment and its information processing capabilities as the focus of analysis, and formal organizational arrangements as the basis of social action" (Kling 1991, 358).

A STIN is not a "thing" to be identified in the world so much as an analytical framework that can be used to understand the relation between technologies, people, and interactions that occur within a given sociotechnical setting. The STIN framework draws both on the social construction of technology approach associated with Wiebe Bijker, Trevor Pinch, and others (Bijker, Hughes, and Pinch 1987; Bijker 1995), and on actor–network theory, which is associated with Bruno Latour, John Law, Michel Callon, and others (Callon 1987; Latour 1987; Law and Hassard 1999). However, the STIN framework focuses on patterns of the routine use of technology as opposed to patterns of adoption and innovation (Meyer 2006). Rob Kling, Geoffrey McKim, and Adam King argue that "the STIN model shares the views of many sociotechnical theories: that technology-in-use and a social world are not separate entities—they co-constitute each other. That is, it is fundamental to STIN modeling that society and technology are seen as 'highly' (but not completely) intertwined" (2003, 54). Similarly, ideas about how new technologies shape the social world also focus on technologies in use (Schroeder 2007b).

This is a good place to make a point about a commonly used word with regard to computer technologies: *users* (we return to this point later in chapters 6 and 7 when we discuss disciplinary differences). The basic problem with this word is that it contains an implicit sense of passive recipients, those who use something created by others. As we have already seen and will continue to see throughout this book, researchers are not generally passive users of technology but are more likely to shape, reshape, design, redesign, and repurpose technologies, even if they are *also* shaped by them. There have been efforts to rethink the concept of users in the human–computer interaction literature (e.g., Cooper and Bowers 1995), science and technology studies literature (e.g., Oudshoorn and Pinch 2005), and information systems research (Lamb and Kling 2003).

Here, we follow the approach taken by Roberta Lamb and Rob Kling, which asks us to reconceptualize "users" as social actors. In this model, "most people who use [information and communication technology] applications utilize multiple applications, in various roles, and as part of their efforts to produce goods and services while interacting with a variety of other people, and often in multiple social contexts" (Lamb and Kling 2003, 197). This view recasts the person engaging with any given configuration of research as active rather than passive (though, again, it can leave room for the shaping of users). This approach reinforces the notion that researchers are not uncritical "users" of technology but actors within professional and personal networks of people and technologies. Their professional domains are of particular interest here as they are bound to shape the way that researchers to engage with new technologies.

As an aside, it is worth mentioning that the term *user* has become ubiquitous with respect to information technologies. For instance, searching publications in the Web of Knowledge for "TOPIC = (users)" shows little use of the word prior to 1960 (total articles = 321), then increases by a factor of two to five times per decade, with the exception of the 1970s, when the number of articles referring to users increased more than 14 times (to 20,458 articles). In the most recent complete decade from 2000 to 2009, more than half a million publications referred to users (total articles = 550,916). Google's NGram Viewer[8] shows a similar pattern in how the word *users* appears in Google's corpus of books, with a distinct upward inflection point around 1960, followed by a steep and steady climb.

Furthermore, the term *user* is problematic because all too often in the world of computer science and technology developers, which is large part of many e-research efforts, the "user" is seen as source of problems and a frequent impediment to the uptake of technological systems. Although it would be difficult to find much academic literature that espouses this view, we have attended many conferences where technology developers bemoan the difficulties of dealing with users: if it weren't for users' inability to recognize the superiority of the technology they have been presented with, the system would work perfectly. Of course, this view is most often put forward when the technology has clearly failed to perform in a way that would enable the "users" to do what they might have hoped to do.

To give just one example: in one meeting we observed, the computer programmers spent much of the time berating the organization that was unhappy with the platform that the programmers had just developed for it. Even though the organization had identified clear objectives, and the platform did not deliver the functions needed to support these objectives, the computer technology partners kept insisting that their platform "worked perfectly" and that it was not their fault if nobody was able to use it or wanted to use it. Their other defense was that the "users" needed to recognize how much more valuable engineer time was compared with user time, so engineers could not be expected to "waste their time" slowing down to the pace required to let those who would actually engage with the platform help improve the interface and functionality; if not for this holdup, the developers would have long since moved on to other projects.

Such an unenlightened view of users is not universal, nor is it the case that those who use technologies do not also inappropriately blame technology developers on occasion. Nevertheless, the term *user* tends to reinforce the idea of "a receiver of what has been provided." Instead, we argue for judicious use of this term and use it mainly in the context of referring specifically to things such as user interfaces, which have a well-understood technical meaning. On other occasions, in referring to those engaging with research technologies in relation to the roles they are performing when they engage with them, we employ the more accurate terms *researchers, scientists, scholars,* and so on. Again, this perspective is integral to understanding STINs and computerization movements.

Computerization Movements and Scientific/Intellectual Movements

Kling was a leading proponent of studying the relationship between adoption of technology and social change, using the concept of "computerization movements" (CMs).[9] In his writing on this topic, he distinguished between social informatics (Kling 1999) and alternative perspectives such as those of sociologists or computer scientists in several ways. In examining how using computer-based systems might transform the social order, Kling (1991) focused on patterns of adoption, use, and social shaping of technology instead of more traditionally on patterns of design and production of technology. By focusing solely on how technology is designed and implemented and on its technological possibilities, Kling cautions, one runs the risk both of espousing naive technological determinism and of falling prey to the hype of computer/information revolutions. A common thread in many of these technology-centric accounts (e.g., Negroponte 1995; Rushkoff 1996; Dyson 1997; and other similar popular "guru" writings) is the assumption that the inherent properties of a given computer-based system will result in positive and predictable types of changes in social behavior. Kling argues to the contrary that

> no one has tried to make a careful case—indicating what kinds of social relations have been transformed, at what level of social activity, under what conditions, and what has not changed. ... As we know from studies of other social revolutions, such as the industrial revolution and the transition from feudalism to capitalism, major social transformations differ in their timing and depth in different places and social sectors. (1991, 346–347)

Kling argues that "what is 'done by computing' is usually affected by a sociotechnical 'system.' ... Usually, there is an interplay of multiple interacting influences" (1992a, 352). To better understand the multiple interacting influences, Kling and Suzanne Iacono examined

> how specialized "computerization movements" (CMs) advance computerization in ways that go beyond the effects of promotion by the industries that produce and sell

computer-based technologies and services. Our main thesis is that computerization movements communicate key ideological beliefs about the links between computerization and a preferred social order which help legitimize computerization for many potential adopters. (1988, 227)

Iacono and Kling (2001) extended this idea to the meaning of technological artifacts at a macrolevel in their discussion of technological action frames, which are similar to the concept of technological frames developed in social construction of technology theory to understand the interactions among the actors of a relevant social group. Technological frames, for social construction of technology theorists, are built up around technological artifacts as interactions among members of a relevant social group converge and move in a similar direction (Bijker 1995). Wiebe Bijker therefore argues that "technological frames" have an affinity to Thomas Kuhn's (1962) "paradigms." In determining the elements of a technological frame, an analyst needs to consider how members of a relevant social group attribute meanings to artifacts and how the artifacts themselves are constituted. Among the elements Bijker (2001) identifies are goals, key problems, problem-solving strategies, system requirements, theories, tacit knowledge, procedures, methods, practice, perceived function, and exemplary artifacts. The importance of the concept of a technological frame is that it allows analysts to incorporate both social and technological elements and to reconcile aspects of purely social constructivist views and technological determinist views of technology: "A technological frame describes the actions and interactions of actors, explaining how they socially construct a technology. But since a technological frame is built up around an artifact and thus incorporates the characteristics of that technology, it also explains the influence of the technical on the social" (Bijker 2001, 15526).

Iacono and Kling (2001) similarly argue that technological action frames are composite understandings of a technology's function and use built up in the language about the technology; these "core ideas" provide potential adopters with a legitimating logic for high levels of investment:

Our main thesis is that computerization movements communicate key ideological beliefs about the favorable links between computerization and a preferred social order which helps legitimate relatively high levels of computing investment for many potential adopters. These ideologies also set adopter's expectations about what they should use computing for and how they should organize access to it. (Kling and Iacono 1995, 121)

Iacono and Kling identify two additional elements in understanding CMs: public discourses and organizational practices. Public discourses are the language—written and spoken public communications—being used to communicate these key ideological beliefs about the technology around which a dominant technological action frame is being constructed to the target audiences of technology adopters. Kling and Iacono (1988) argue that to understand CMs, one needs to understand not only the participants and their social organization but also their relations with, among other things, the media. Kling (1992a) further argues that the optimistic claims made about technology in the popular press and within organizations can be partially understood as an understandable effort to garner scarce resources for technology acquisition. Finally, organizational practices are "the ways in which individuals and organizations put technological action frames and discourses into practice as they implement and use technologies in their micro-social contexts" (Iacono and Kling 2001, 100). This stress on organizations was integral to Kling's work.

One question that can be asked here is whether e-research constitutes a CM at all? Christine Hine is among those who discuss e-science as a CM and contends that the concept of CMs offers normative potential in the realm of e-science: "If much of the potential of a computerization movement is in stimulating creative reactions against it, then more could usefully be made of occasions for expressing doubt and disaffection. This kind of debate should be encouraged" (2006, 45). Based on her examination of a 2002 UK government report (Select Committee on Science and Technology 2002), she suggests that CMs—in this case, the effort "to assess the current state of systematics in Britain, with particular reference to its capacity to fulfill the government's commitments under the Convention on Biological Diversity"—can themselves be resources leveraged to mobilize support and funding for new technologies (2006, 32, 42–43). To be sure, such mobilizations are part of how researchers obtain support for their efforts. However, as we shall see, this is only one aspect of what drives e-research; the analysis provided by Iacono and Kling can also take us beyond this resource-mobilization perspective.

Noriko Hara and Howard Rosenbaum also argue that e-science constitutes a CM: "e-science ... is a fairly constrained CM that encourages discourses among a specialized population. ... Though the scope of e-science is rather small, it is a CM because technological action frames have arisen around it, it has engendered various types of discourses, and it involves a range of computerization practices within a range of organizations" (2008, 234). They build on the original conceptualization of CMs and suggest that a number of subtypes more accurately reflect the range of diversity displayed in CMs. Using their typology, e-science/e-research would be a bundled, nonmarket-driven, internal, narrow, positive CM. Other CMs with a similar cluster of characteristics include expert systems, paperless courtrooms, virtual reality, and digital libraries.

As observers of e-research and in particular part of the UK e–social science community, however, we consider e-research only in part a CM, championed by a number of organizations and actors. It is also a wider, often unconscious effort to introduce digital technologies into the practice of science, social science, arts, and the humanities. Scott Frickel and Neil Gross have argued that scientific/intellectual movements (SIMs) emerge under certain social conditions as competing research traditions seek to "gain adherents, win intellectual prestige, and ultimately acquire some level of institutional stability" (2005, 205). One of the elements of a SIM that Frickel and Gross focus on is that although SIMs can eventually enter mainstream intellectual thought, a movement is a SIM at least in part because "at the time of its emergence, it significantly challenges received wisdom or dominant ways of approaching some problem or issue and thus encounters resistance. Gradual or incremental changes—the stuff of Kuhn's (1962) 'normal science'—are best understood from the standpoint of institutional drift. By contrast, SIMs involve dramatic breaks with past practices" (2005, 207).

One question is whether e-research is more accurately understood as a CM (which at its core is about communicating and demonstrating the desirability of applying computation to a domain) or as a SIM (which is about communicating how certain intellectual positions or scientific approaches to a phenomenon will better advance science or, here, how the intensive adoption of digital research technologies and data will drive science and knowledge for the better) or a combination of both CM and SIM. In any event, bringing reluctant users in, pushing for widespread adoption of new research technologies, and learning how these technologies strengthen communities and organizations are parts of the same process. As we have started to argue, e-research combines elements of a CM and a SIM, but these elements advance knowledge via hardening cores of sociotechnical systems that enable knowledge advance at the research front. The aggregation of these shifts can only be discerned as a fluid—but increasingly networked and digital—change in the landscape of research and advancing knowledge, a landscape that we describe in the next chapter.

e-Research in Society and in Academic Disciplines

When we talk about e-research and computational approaches to research, we are at some level talking about the algorithms that not only turn data about the world into the ones and zeros of binary computing but perform operations on those ones and zeros using logic and structure. Algorithms are machines performing these operations by following mathematical models and instructions (Moschovakis 2001). As software applications become more and more complex, individuals' ability to understand the entire workings of the application implementing the algorithms diminishes, particularly among those who did not participate in the creation of the algorithms and their implementations in software. The complexity of algorithms that are increasingly affecting public life has been discussed in some contexts (such as the increasing reliance of Wall Street trading on algorithms; see Salmon and Stoke 2010), but more and more domains are becoming at least partly controlled by algorithms. Automated systems that suggest books to buy based on your choices and the choices of people like you, satellite navigation systems suggesting how to drive from one place to another, advertisements placed in web pages based on one's browser history, and a multitude of other examples proliferate. This is also true in research. Scientific code that is hundreds of thousands of lines of code long, written by teams of people, and essentially impossible for a single human to read and check are increasingly being used to process and analyze data. What is the impact of this increasing use of complex algorithms on science? We pursue this issue in chapter 10.

One reason for starting to sketch our broader argument is that it allows us to see why neither Internet studies nor the sociology of science and technology has been able to address the transformations we have charted here, which should be central to the endeavor of explaining how knowledge is changing. This is because neither has the conceptual tools to grapple with how knowledge is changing across various domains. Internet studies is not focused on science and knowledge, and the sociology of science and technology has dealt with individual case studies of e-research. Here we are interested in certain ways of doing science and research by means of digital tools and data—at base, a mathematization of knowledge or a means of producing knowledge by using computer algorithm to manipulate data—across the range of academic research. Individual case studies do not allow comparison of how research communities are oriented to different shared tools or objects (here, objects represented in digital form and computationally manipulated) and how scholarly practices are therefore becoming transformed. Yet for the other disciplines that we mentioned earlier—including media studies, political-institutional approaches, economics of innovation, and computer science fields—understanding how science and knowledge are becoming transformed and how they advance at the research front is not a core concern, even if it brings valuable perspectives to the topic of e-research. Hence, we have tried to put forward ideas that place the transformations of knowledge at the center but also, given an account of the technological cores around which communities focus, that describe how these transformations push research in new directions by adopting new research technologies and ultimately how these instruments create new sociotechnological systems on which large swathes of research rely. We continue to develop this theme throughout the book, even though this task is bound to remain incomplete because it entails an understanding of a constantly ongoing process whereby various frontiers are advancing around computational techniques for manipulating digital data. In understanding the transformations that are taking place, we identify the commonalities across various fields and how the foci and advances of research communities can be generically understood from a social science perspective. We hope that this endeavor can bring different disciplines studying the Internet and research together around a shared object: transformations in knowledge. Whether such a shared object can overcome the fragmentation among social science disciplines remains to be seen.

As we demonstrate throughout this book, the reconfiguration of disciplines and the globalization of knowledge is indeed a feature of e-science, but skeptics have a point insofar as these changes need to be demonstrated in practice rather than merely asserted. To do this, we can further develop the concept of research technologies and give an account of how e-science tools exemplify this concept. We have begun to do this in this chapter, but now, having added the ideas that movements attempt to push and instantiate new forms of computerization and that the focus should be on technologies in use, we can turn to looking at how computational approaches to research have grown over the long term.

3   The Rise of Digital Research

In chapter 1, we gave examples of e-research and argued that this type of research is becoming more visible. In chapter 2, we provided an account of how e-research is transforming knowledge and how this "movement" is spreading to new domains. In this chapter, we chart this rise. Over the past decade, various funding programs have enthusiastically promoted e-research in the sciences, social sciences, and humanities. The so-called Atkins report (Atkins, Droegemeier, Feldman, et al. 2003) was perhaps the most well-known document produced by such programs, laying out plans for a "cyberinfrastructure," but the report by Francine Berman and Henry Brady (2005) also laid out an ambitious vision of e-research in the social sciences. The latter report makes the case that the social sciences have the potential to make bold advances in the near future, including the federation of online data sets, experimentation in human behavior with virtual worlds, and the collection and analysis of data from the extensive digital traces that people produce. This chapter's aim is to present empirical data that will allow us to gauge the extent to which this vision is being realized.

Some initial studies to gauge the diffusion of e-science and e–social science have examined the adoption and diffusion of different social science digital research tools, as was done in a survey for the former UK National Centre for e–Social Science (NCeSS) (Dutton and Meyer 2009; Meyer and Dutton 2009) and in a survey for the Accelerating Transition to Virtual Research Organization in Social Science (AVROSS) project, funded by the European Commission (Barjak, Wiegand, Lane, et al. 2008). But the notion of simple diffusion (see Rogers 2003) provides a limited perspective without providing the broader context of how research technologies are critical to the *advance* of the disciplines, as opposed to simply allowing the same research to be done faster, more conveniently, or at a larger scale: it might be, for example, that certain tools or data sources are widely used but have a limited or incremental impact on the types of research questions scholars are able to address. One example is online surveys. Although online surveys do raise some new methodological issues for practitioners (Dillman 2000; Vehovar and Manfreda 2008), the basic types of research questions addressed by online surveys are not remarkably different from those that can be answered using paper or telephone surveys. Online surveys are clearly faster to deploy, less expensive to administer, and more convenient to run, but they arguably are not advancing the social sciences in ways that would otherwise have been impossible.

It is possible to gauge e-research in a number of ways. One is to use scientometric measures, such as those reported later in this chapter, to understand the publications overtly discussing e-research and related topics. With this approach, we focus on the role, scale, and nature of e-research in the generation of knowledge, particularly in relation to the fields and disciplines engaged in e-research. But, first, we look at funding and the growth of e-research.

Funding e-Research

It is clear that two e-science programs were pioneers—those of the United States and the United Kingdom.[1] Programs in the European Union, Sweden, and Germany (among the examples discussed here) and many others have followed. In the United Kingdom and the United States, large-scale initiatives began around 2000. In the United Kingdom, a £250 million, five-year e-science program was initiated in 2001 in order to develop tools, technologies, and infrastructure to support multidisciplinary and distributed collaborations. The program tried to address the so-called data deluge (Hey and Trefethen 2003) in the physical and biomedical sciences. The UK e-science program was conceived of as a collection of pilot e-science projects in a range of disciplines underpinned by a "core program" concerned with generic e-science infrastructure (Hey 2004).

In the United States, the label *collaboratories* was in use for some time before the cyberinfrastructure program was developed (Finholt 2003), although, as mentioned, the Atkins report (Atkins, Droegemeier, Feldman, et al. 2003) led to a US program spearheaded by funding from the National Science Foundation (NSF). The NSF created the central Office of Cyberinfrastructure (OCI) in August 2005, and the vision was extended to the social sciences in the report by Berman and Brady (2005). In Germany, e-research started relatively late, but more than a dozen projects began in the second half of the first decade of the 2000s, and Germany also had a strong presence in European Union–funded e-science projects. In Germany, the main funding has come from the German Federal Ministry of Education and Research—rather than the German Research Foundation, which is the main academic research funding body in Germany. This alignment couples the German program more closely with the national system's broader aims of innovation and training—in contrast with the UK and US focus on academic research.

One illustration of how e-science programs can be shaped by national politics is the debate in German e-research about how the individual states—the *Laender* in Germany's federal system—should contribute when the capacity of high-performance computing resources in one state is tapped by other states, an issue that has also played out on a European level. Another issue in research policy is where efforts should be focused. In Germany, there has been a strong emphasis on the physical science and engineering disciplines as well as on the humanities, with comparatively little emphasis on social science—unlike in the United States and the United Kingdom.

In the United Kingdom, e-research efforts were largely organized under umbrella organizations in the different fields: the National Grid Service, the National Centre for e-Science, NCeSS, the Arts and Humanities e-Science Support Centre, and the Digital Curation Centre. Note also, however, that nearly all of these efforts have either been abandoned as stand-alone programs or transformed over the course of time.

Funding data worldwide is extremely fragmentary, and it is difficult to determine the exact levels of overall funding dedicated to e-research. This difficulty is due in part to varying reporting systems worldwide, but also to lack of a reported connection to e-research. Funding for cyberinfrastructure projects, for example, may be channeled through an office or directorate dedicated to funding e-research or through a standard funding body, with little indication in searchable databases that these projects are dedicated to building e-research capacity. For instance, two large cyberinfrastructure projects funded by the NSF were quite prominently announced on the NSF OCI website, but in the NSF database they are linked not to the OCI directorate but to the Division of Behavioral and Cognitive Sciences. Nevertheless, just these two well-publicized projects alone—Bennett Bertenthal's project at Chicago to develop the Social Informatics Data Grid and Michael Macy's project at Cornell to build tools for working with Internet archive data—are an indication of the large resources dedicated to e–social science, each project being funded for approximately US$2 million. Within the OCI itself, of course, a large number of projects have been awarded: according to the NSF Database,[2] the OCI funded 342 projects in the period 2000–2008 for a total of more than US$375 million, distributed among 158 different organizations. But these organizations are continuing to undergo changes, as has happened in the United Kingdom, so that the OCI, for example, has been folded into the larger and long-established Directorate for Computer and Information Science and Engineering at the NSF. Other organizations, spread across a number of funding agencies, have also funded e-research projects in the United States.

Franz Barjak, Gordon Wiegand, Julia Lane, and their colleagues (2008) examined funding by region for social science e-infrastructure projects and found that, in general, continental Europe and the United States funded projects at a higher level than the United Kingdom funded similar projects. The average project size in the AVROSS sample, for instance, was €734,000 in the European Union and €522,000 in the United States, but only €307,000 in the United Kingdom. All three amounts funded, however, are considerably higher than funding from other countries in the sample, which averaged only €160,000. There is also considerable cross-discipline variation: the average total budget for the linguistics projects these authors examined was €776,000, but other fields averaged less than half that (with economics, social geography, and sociology all averaging €300,000, plus or minus €4,000); one exception was archaeology, which averaged the lowest of any field, with only €148,000. In the United Kingdom, the several projects funded specifically within the NCeSS program have also been funded on a generous scale compared with other UK social science programs, garnering at least £20 million over the length of the program.

Mapping e-Research Output

The data from the studies mentioned earlier are aimed at both giving an indication of how e-research technologies are being adopted (or not) within different fields and understanding what barriers and enablers there are to adoption. This approach gauges the e-research domain focused on the development and production of scientific capacity. Another approach examines scholarly outputs as traces for understanding how computational approaches to research more generally are reflected in publications (here we can also recall the discussion about the "visibility" of e-research at the end of chapter 1).

Taking the latter approach, we created a data set of publications from the Scopus database on the topic of e-research, broadly defined. The search string for this study was designed to retrieve a wide variety of articles on e-research, including terms and variants anywhere in the titles, abstracts,

The Rise of Digital Research

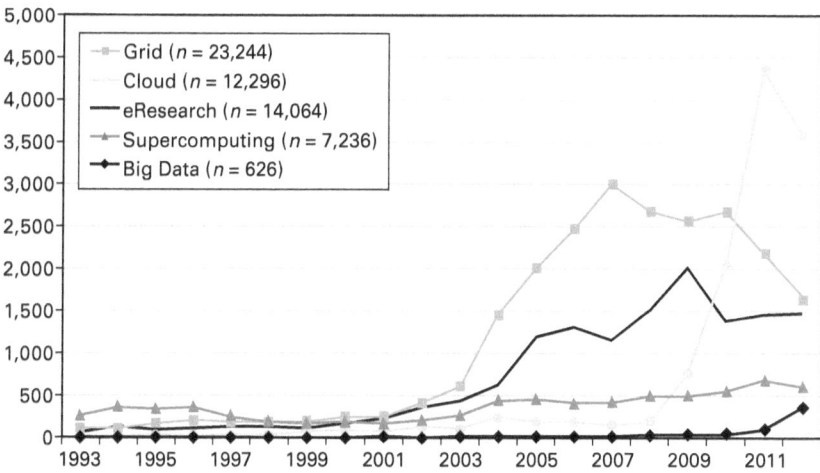

Figure 3.1
Publications on collaborative computing topics, 1993–2012.
Source: Data compiled from Scopus.

or keywords of the Scopus record related to e-research, e-science, e-humanities, e–social science, cyberscience, cyberinfrastructure, cyberresearch, e-infrastructure, digital humanities, humanities computing, social science infrastructure, humanities infrastructure, social science grids, and big data.[3] The search yielded a total of 14,064 articles from 1993 to 2012, including 4,738 journal articles, 7,962 conference papers, and the remainder a variety of other academic outputs (primarily editorials and reviews).

To put these results in context, we compared several other topics that are also related to computation and computational infrastructure. In figure 3.1, we see that the results tell a story that supports our argument that this area has grown: even though topics relating to computation in research have been part of scholarly publishing for many decades (going back much further than we show in this graph), an inflection point indicates considerable growth of publications on these topics in the 2003–2004 period. Works on Grid computing,[4] for instance, more than doubled, from 238 publications in 2001 to 604 in 2003; then it more than doubled again in just one year, increasing to 1,459 publications in 2004. By 2007, there was a peak of 2,995 publications on Grid computing, before falling somewhat in recent years. However, even with this relative decline, publications about Grid computing are most prominent overall for this time period, with a total of more than 23,000 publications. Of the topics we are considering here, only Cloud computing surpassed Grid computing at any point over the past decade and not until 2011.

Cloud computing[5] publications began to rise as those on Grid computing began to tail off somewhat, which is consistent with what was happening on the ground in the e-research and e-science communities. Whereas the Grid was a central focus of early e-science efforts (see, for instance, Foster and Kesselman 2004), during the course of the first decade of the twenty-first century, it became clear that although the Grid worked very well for certain disciplinary communities and applications such as physics (see, for example Zheng, Venters, and Cornford 2011; Pearce and Venters 2013), there were fewer clear applications of Grid technologies and greater training and infrastructural barriers to using the Grid for other disciplines, including the social sciences and humanities. A steep rise in publications related to Cloud computing occurred after 2008, when such publications increased from 178 in 2008 to 757 in 2009 and then to a high point of 4,324 in 2011. One of the factors that influenced this increase is that whereas Grid computing was a specialist resource with relatively few applications outside the science and research domain, Cloud computing is a commodity technology widely used in industry and business as well as in research applications. In the popular press, there are literally tens of thousands of articles per year on Cloud computing; in just the first week of November 2012, for instance, 2,592 articles in the English-language media[6] mentioned Cloud computing.

This brings up an interesting point to consider when discussing research uses of technology: certain technologies either transfer out of the research domain into the broader public and business world (such as the Internet) or start largely in the commercial or broader public sectors but are appropriated for research (such as Cloud computing and big data). As we saw in chapter 2, this flexible appropriation in various domains is precisely one of the features of research technologies that makes for their spread, according to Terry Shinn and Bernward Joerges (2002). However, such flexibility also makes it more difficult for scholars of science and research to disentangle these technologies when trying to understand their uses in research as opposed to their public and commercial uses. The research uses are frequently dwarfed by uses in the commercial sphere or the public sphere. Thus, for instance, knowing what proportion of the growth in academic research mentioning Cloud computing is due to the widespread growth of Cloud resources for business and public is impossible.

This will also potentially be a problem for another recently growing topic related to e-research: big data.[7] We discuss big data later in the book

The Rise of Digital Research

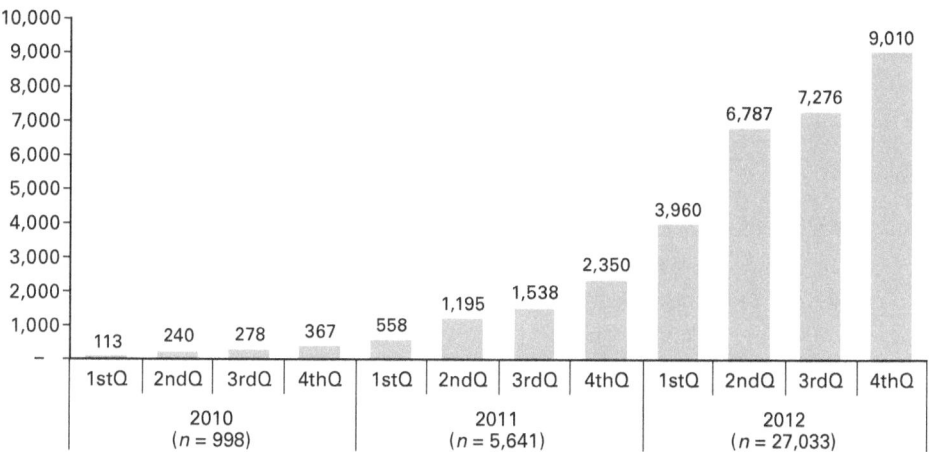

Figure 3.2
Number of news articles on big data.
Source: Data compiled from Nexis.

(chapter 6), but for now it is enough to note that the phenomenon of big data began to sweep across certain academic disciplines and many businesses in 2011 and 2012. Although academic publications on the topic of big data show an uptick in 2012 (to 353 articles from 87 in 2011), these publications are swamped by the number of news articles on big data in the global English-language press, where more than 27,000 articles mentioned big data in 2012 (see figure 3.2).

This public framing of a topic contributes to the technological action frame that is built as part of a CM. However, the knife can cut both ways: although widespread public interest in a phenomenon such as big data can make it easier for researchers to explain their interest in what might otherwise be seen as an esoteric topic, it can also have the opposite effect when discussion of the topic comes to be seen as part of a hype cycle and people grow disillusioned with the promises of new technology (Fenn and Raskino 2008). The fluctuation may also occur because people are uncertain about what a new technology can actually do (MacKenzie 1999) or because those closest to the technology start to see the problems inherent in the application of a complex technology that was promised to be simple (Dutton and Shepherd 2006; Dutton and Meyer 2009).

Turning to one of the longest-standing computational research topics in our comparison in figure 3.1, we can see that publications related to supercomputing[8] saw a less dramatic increase during the same time period as other topics but nevertheless experienced steady growth from a fairly steady rate of 200–300 publications per year throughout the 1990s to a high of 678 publications in 2011.

Finally, we can look at our main topic of interest in figure 3.1, publications related to e-research. Here we see steady overall growth since 2003, with some annual variation, but an overall upward trend. The e-research sample is our main interest here, so we analyze these data in more depth.

In figure 3.3, we see the overall growth in publications on e-research and what proportion of publications came from various disciplinary areas. From a few scattered articles in the late 1990s, there is steady growth from 2000 on, with particularly notable increases in the 2004–2005 period. Each full bar shown in the figure represents the entire data set of e-research publications, and the shorter, shaded portion of each bar shows which proportion of the publications span multiple broad fields (i.e., computing/engineering and science or science and social science). The proportion of publications that span multiple fields averages just more than half (53 percent) across the data set, with some annual fluctuation.

As is apparent in figure 3.3, computing and engineering is central to e-research. This finding is unsurprising at one level because e-research is at its core concerned with bringing computational approaches to the sciences, social sciences, and humanities. It is also understandable that the method we have used to discover publications in this area (which inevitably favors publications that *discuss* computational approaches to research over those that may *use* such methods but do not explicitly talk about them in a discoverable way) might be somewhat biased toward publications with some computer science or engineering involvement. What is surprising, however, is the overwhelming extent to which computer science and engineering are central to what has been happening in e-research over the past decades. We made this point in an earlier publication, where we noted similar trends and argued that "computer science has not just been the lucky recipient of the attentions of interested e-researchers, but that many of the central actors pushing e-research forward ... are based in computer science" (Meyer and Schroeder 2009a, 256). In fact, if we look at the top-25 authors in our data set in terms of outputs, 21 are computer scientists, with the remainder in engineering (2), informatics (1), and chemistry (1).

The bottom portion of figure 3.3 shows the total times (by the end of 2012) that the items in the data set have been cited, broken down by year of original publication. Although these publications have been cited nearly 80,000 times, we can see that the ones written near the beginning of the concentrated e-science/cyberinfrastructure efforts in the United States and

The Rise of Digital Research

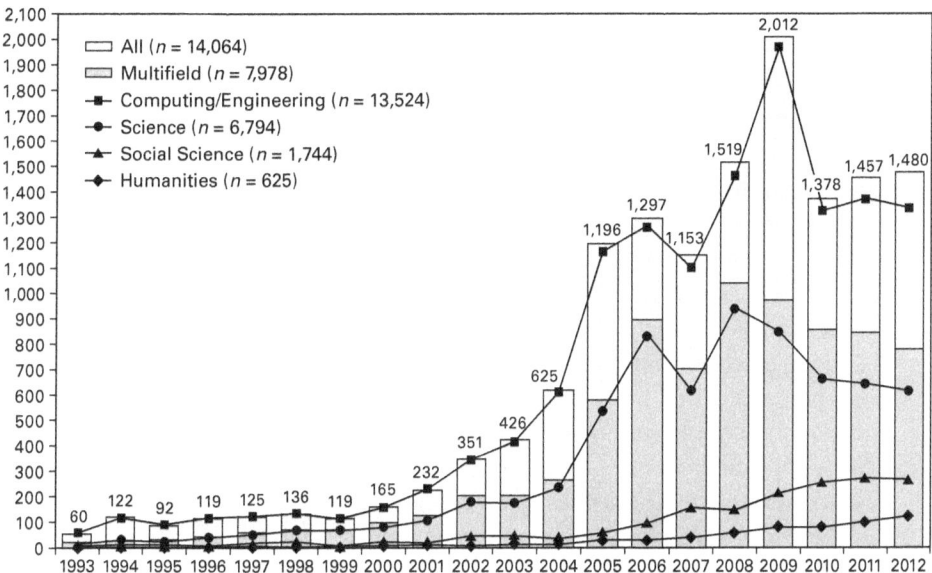

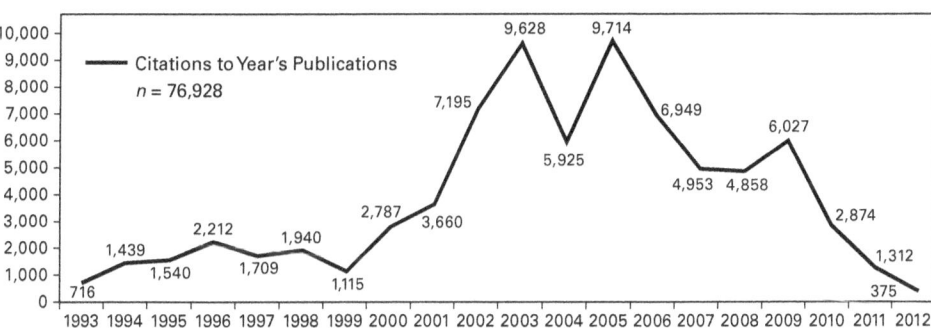

Figure 3.3
e-Research publications and citations, 1993–2012.
Source: Data compiled from Scopus.

United Kingdom (in the 2002–2006 period) have been cited the most. Part of the decline in recent years is due to the fact that a certain amount of time is required before citations of articles can work their way into publication themselves. It is also worth noting that although the 2003 data are distorted by the extremely high citation of one article in particular (a bioinformatics article cited more than 3,300 times), this is not true of any of the other years, when several articles are cited hundreds of times, but none cited more than a thousand times.

Turning our attention to the science, social science, and humanities data, we can see that scientific fields (which include the physical, natural, and life sciences) are far more prominent in the data set than the social sciences (including business and economics) and the humanities: publications in science fields make up nearly half of the overall data set, whereas the publications in the social sciences compose just more than 10 percent, and those in the humanities make up less than 5 percent.[9] We return to the topic of disciplines in later chapters, but in the context of these data we can see that even though the social sciences and humanities have been the target of considerable attention in terms of funding streams, dedicated conferences, and active recruitment of participants discussed elsewhere in this book, they have lagged behind the sciences in terms of bringing computational approaches into their research and actively discussing this fact in their publications. We should also note, however, that humanities data in particular has less complete coverage in the Scopus database, so other search techniques and other databases might yield a bigger absolute sample.

The publications represented in this data set are far from the sum total of durable knowledge generated on the topic of e-research over recent decades. In fact, the gray literature (which includes nonjournal publications such as reports, dissertations/theses, unpublished conference proceedings, and working papers) contains what appears to be at least twice as many publications as those in journals, but this gray literature is very difficult to extract and analyze in any meaningful way because it is spread across many websites and has inconsistent or nonexistent metadata. For instance, using the same default search term on topics related to e-research on Scopus mentioned earlier retrieves 14,064 raw documents but then also shows that an additional 19,796 references (not among the original set) can be extracted from those publications' references. These additional 20,000 documents include a number of conference proceedings that are not formally published as well as other gray literature. In addition, Scopus shows an additional 1,127,302 documents in its scholarly web search engine[10] (which includes items such as web pages and other sorts of gray literature held in university repositories). Among these more than one million items, there are more than 1,000 books, nearly 10,000 theses and dissertations, and more than 30,000 items held in academic digital repositories. These publications are not only difficult to analyze, of course but also difficult for researchers themselves to discover, read, cite, and critique. So although our main sample of just more than 14,000 articles can be said to form a core published corpus on e-research, it is only the tip of the iceberg in terms of all the words that have been written on the topic over the past few decades. And it is worth stressing again that publications with explicit mentions of e-research and its cognates form a small subset of research that falls under our definition of e-research (and it is likely that some publications on e-research and cognates do not fall within our definition).

To give an example of the extent to which gray literature and differing database construction techniques alter the scientometric view of a topic, we can look at some data from a recent project of ours in the realm of digital humanities. In this project[11] with the Bodleian Libraries of the University of Oxford, we were interested in the usage and impact of the Early English Books Online (EEBO) resource in general, but more particularly in the Text Creation Partnership (TCP) resource, the EEBO-TCP, which has been creating the full-text transcriptions of the books in the EEBO collection. EEBO is a collection of all works printed in England or English from 1473 to 1700 CE, and the digitized pages are available to subscribers via ProQuest;[12] EEBO-TCP creates searchable XML-encoded digital editions of the EEBO texts and provides them through a separate interface[13] that is linked to the commercial product. These texts have been digitized and encoded over a period of 15 years; the text encoding has cost about $20 million and has taken more than 90 person-years of effort.

In the EEBO-TCP study, we looked for publications that mention using EEBO or EEBO-TCP anywhere in the text or references and found very different results using different databases. Scopus, which focuses on journal publications, found 239 publications, and JSTOR, which is a more specialist humanities resource, found 299 publications. Google Scholar, however, which indexes a far greater variety of published and nonpublished sources, found 10 times as many publications (3,450). Looking for theses and dissertations that cite or mention EEBO/EEBO-TCP, ProQuest Dissertations & Theses found 458 publications, and Scopus Theses & Dissertations found 1,875 (Siefring and Meyer 2013).

This large variation based on the size and scope of each database is complicated even further when one considers citation habits with regard

Table 3.1
Publications Related to Early English Books Online

Database	Number of Publications Citing EEBO or EEBO-TCP
Google Scholar[a]	3,450
JSTOR[b]	299
Scopus[c]	239
Scopus Theses & Dissertations	1,875
ProQuest Dissertations & Theses[d]	458

a. Google Scholar search term: "eebo-tcp" OR "eebo tcp" OR eebo OR "early english books online."
b. JSTOR search term: eebo-tcp OR "eebo tcp" OR eebo OR "early english books online" in full text, including all content.
c. Scopus search term: ALL ("eebo-tcp" OR "eebo tcp" OR eebo OR "early english books online").
d. ProQuest Dissertations & Theses: The Humanities and Social Sciences Collection search term: "eebo-tcp" OR "eebo tcp" OR eebo OR "early english books online" in full text.
Source: Siefring and Meyer 2013.

to digital resources. In the same study, we surveyed EEBO users ($n = 172$) about their citation habits, and only about half of the researchers said they would include any indication of their use of these digital resources in their publications. Even though few relied on printed versions of documents for their research anymore, many still cited the printed versions as if they had consulted them (34 percent) or included only an indication that they had used an online version, but not a URL or specific resource name (6 percent). Forty-five percent of users in this survey included citations to both the print version and a URL, a rate that is consistent with the findings of earlier work examining users of various resources. In that study (Meyer, Eccles, Thelwall, et al. 2009), we found the likelihood of the inclusion of any reference to the digital resources being consulted varied by database and that this variation was in part due to the affordances of the interface (e.g., providing a "suggested citation" on each page that included a short URL increased the likelihood of the digital resource being cited) and in part to different disciplinary citation practices. Further, in that study only 9 percent of users of Histpop—the Online Historical Population Reports[14] (for UK historical population census data)—cited just the printed version, 55 percent cited the original plus a URL, and 36 percent cited just the online version (Histpop gives a suggested citation that includes the URL). The

British Library Newspapers data in the study, however, showed 53 percent of users cited just the printed version of the newspaper, with no indication they had consulted the digital resource, even though 56 percent of the same users said they had never used the original printed version of the documents available through the digital site (Meyer, Eccles, Thelwall, et al. 2009).

The point of this brief digression is to make clear that when we are discussing how publications are reflecting the move toward computational approaches to research, we have to keep in mind that not all of the e-research activity is discoverable using these techniques and that in some fields there are biases against mentioning digital resources. One humanities scholar interviewed in our research said, "I do feel pressure to work more with originals than with the digital images, but for the most part I do feel like I get more out of using these images on my computer. But there's a certain pressure that that's not what top scholars do because that's not what top scholars did 25 years ago" (historian, personal interview, September 2010).

Turning our attention back to our sample of publications, we also find it interesting to know where this work on computational approaches to science and research is taking place geographically. To determine this, we extracted the names of all the individual authors of the 14,064 papers in our sample and extracted each author's country from his or her address where possible (13,278 publications had sufficient information to extract country-specific data and allowed us to determine the main author, defined as the corresponding author[15] for the publication, where identified; in tables 3.2–3.4, we further limited the sample to those publications from the most recent decade, resulting in a total of 11,766 publications with author and location information).

In table 3.2, we see the authorship of e-research articles by continent, focusing just on the most recent decade, during which the bulk of the publications were written (11,766 of 13,278, or 88 percent). We extracted 39,622 total author locations from the data (although, of course, some authors appear in the data set multiple times, a factor we discuss later). We can see that Europe and North America are predominant, as one might expect in English-language publications, but Asia is a close third. Oceania (primarily Australia and New Zealand), South America, and Africa are each an order of magnitude less prominent. Again, this is not a particularly surprising finding, particularly for South America and Africa, given what is already known about the unequal global distribution of publications and academic resources. Nevertheless, the findings point to something we see

Table 3.2
e-Research Authorship by Continent, 2003–2012

Continent	All authors	Main author	Multiple Authors		Multiple Countries		Multiple Continents	
			n	%	n	%	n	%
Europe	13,477	3,906	3,169	81.1	1,006	25.8	432	11.1
North America	13,468	3,867	2,995	77.5	529	13.7	438	11.3
Asia	10,798	3,399	3,057	89.9	407	12.0	308	9.1
Oceania	998	312	247	79.2	69	22.1	67	21.5
South America	673	197	175	88.8	34	17.3	33	16.8
Africa	208	85	66	77.6	23	27.1	22	25.9
Total	39,622	11,766	9,709	82.5	2,068	17.6	1,300	11.0

Source: Data queried from Scopus.

time and again: although computerized tools are often discussed as democratizing influences, breaking down barriers of access and participation, in reality the structural barriers to meaningful contributions in science and research remain in place for scholars working in the Global South. If we look at individual authors, 11 of the 25 most prolific authors on e-research as measured by total number of publications are based in the United States, 7 are based in the United Kingdom, and the remainder are from Australia (3), Germany (2), Poland (1), and the Netherlands (1).

This point about inequality in the production of knowledge is worth visiting briefly. As argued in a study on publications in the physics preprint repository arXiv.org,

> It is important to underline the differentiation between access and participation. Providing access to electronic forums is relatively simple: develop a web site with suitable features, submit the URL to search engines, list-servs, and friends, and wait for people to show up and access the available information. Participation is an entirely different dimension of behavior. To participate actively in an electronic forum such as arXiv.org, one cannot just lurk and browse or read other people's research manuscripts. People also must engage in the behavior necessary to have a voice in the online information resource in question. In the case of arXiv.org, scientists must also post their own research manuscripts for others to read (and discuss). (Meyer and Kling 2002, 2)

Thus, although we have no way of knowing from our sample if the topics related to computational approaches to science and research are

Table 3.3
Countries and Citation Patterns of e-Research, 2003–2012

Top 15 Countries	Authors (n)	Publications (n)	% Cited	Average Citations	Average Number of Authors	% Multiauthor	% Multicountry	% Multicontinent
United States	12,452	3,523	50.4	12.42	3.53	77.4	12.4	10.6
China	5,882	1,820	31.2	4.41	3.26	92.4	8.8	6.6
United Kingdom	3,741	983	61.4	9.28	4.03	81.6	23.8	11.8
Germany	1,945	562	56.8	6.24	3.42	76.5	25.3	9.6
Japan	1,287	379	42.7	5.60	3.35	82.8	15.3	10.8
Italy	1,264	346	63.0	7.11	3.62	85.5	23.7	11.0
France	1,089	352	64.2	6.94	3.07	81.0	30.1	13.1
Spain	1,065	284	51.8	30.53	3.85	91.9	25.7	12.0
South Korea	1,006	287	33.4	3.27	3.54	91.6	7.7	5.6
Australia	929	283	48.1	8.29	3.27	80.6	20.8	20.1
Canada	834	290	53.4	6.78	2.71	76.2	25.5	16.9
Netherlands	693	207	59.9	7.52	3.35	86.0	34.3	16.4
Poland	601	189	59.8	3.79	2.86	78.3	14.3	4.8
India	578	206	35.9	5.11	2.78	89.8	11.2	10.7
Brazil	565	164	47.0	5.36	3.38	90.9	14.6	14.0
Total*	39,622	11,766	48.6	9.20	3.26	82.5	17.6	11.0

* Including countries not displayed (106 total countries in data set). All averages and percentages are based on corresponding authors.

Table 3.4
Twenty-Five Prominent Institutions Publishing on e-Research

Institution	Number of Publications
Indiana University	156
University of Manchester	140
University of Illinois at Urbana-Champaign	136
University of California, San Diego	122
University of Oxford	112
Chinese Academy of Sciences	103
Purdue University	102
University College London	97
University of Amsterdam	95
Argonne National Laboratory	95
Tsinghua University	94
University of California, Berkeley	89
University of Southampton	88
University of Edinburgh	86
Massachusetts Institute of Technology	83
University of Chicago	82
AGH University of Science and Technology, Krakow	81
Oak Ridge National Laboratory	79
Lawrence Berkeley National Laboratory	73
San Diego Supercomputer Center	72
Carnegie Mellon University	72
University of Texas, Austin	69
University of Melbourne	68
Pennsylvania State University	67
Shanghai Jiaotong University	67

influencing researchers in the Global South, we can with certainty say that they are not widely influencing the future directions of e-research there via publications. As Espen Aarseth puts it, "A reader, however strongly engaged in the unfolding of a narrative, is powerless. Like a spectator at a soccer game, he may speculate, conjecture, extrapolate, even shout abuse, but he is not a player" (1997, 4).

Table 3.2 also provides valuable information about coauthorship in our sample. It can be seen that almost 83 percent of the papers in this sample are coauthored, which is consistent with other data about the growing role of teams in the production of knowledge (Wuchty, Jones, and Uzzi 2007). Furthermore, almost 18 percent of the sample has authors from multiple countries and 11 percent from multiple continents—again, data that are consistent with previous work looking at international collaboration (Wagner and Leydesdorff 2005). Unsurprisingly, European authors are the most likely to collaborate with authors in other countries but are equally likely

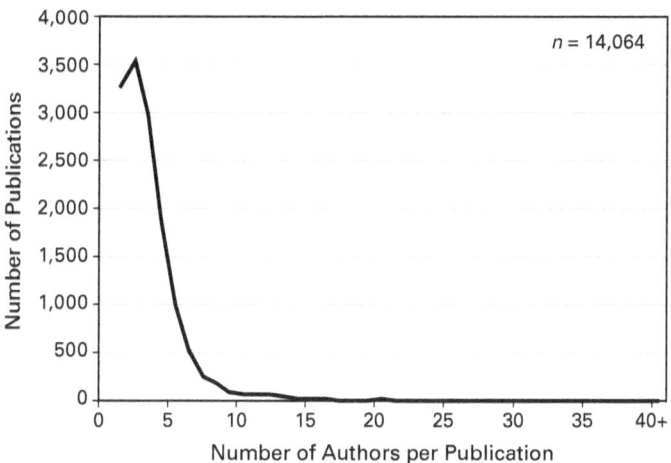

Figure 3.4
Authors per publication.
Source: Data compiled from Scopus.

as their North American counterparts to collaborate with authors on other continents. We suspect that if the North American data could be broken down by state, it would show a similar pattern of cross-border collaboration as is seen in Europe.

In figure 3.4, we see a simple distribution of authors per paper in our e-research sample, which follows a fairly typical long-tail distribution with one small exception: publications with two authors are slightly more common than single-author publications. Again, this is what one would expect in light of the earlier data showing more than 80 percent coauthorship, but it highlights an element related to both e-research and academic publishing in general: the stereotype of the lone scholar is increasingly a relic of the past.

Turning back to the geographic distribution of authorship, we can also look at country-level data aggregated across the data set to find the countries most represented in publications on e-research and at the author and affiliation data for the most prolific authors and institutions in the data.

If we look at the countries in the data set (see table 3.3), the United States, China, and the United Kingdom are the most prominent. However, although China surpasses the United Kingdom in the volume of publications on e-research, their impact per publication is much lower, with less than one-third of Chinese publications cited by other publications, with each cited only 4.4 times, compared with nearly two-thirds of UK papers being cited, with each cited an average of 9.3 times. South Korea and India also perform poorly in terms of the percentage of their publications on e-research that are cited (33 and 36 percent, respectively) and correspondingly have low impact in terms of the number of times each paper is cited as well (3.3 and 5.1). This begs the question of what is going wrong in terms of publication impact in Asia and what is going right in Europe. Language seems insufficient to explain the difference, considering that Spain has the highest number of citations per publication (30.5), and Italy and France have the highest proportion of their publications cited (approximately two-thirds in both cases).

There are also a few notable differences in collaboration patterns at the country level. First, the single-author model, although not prominent anywhere in this sample, is more common in the United States, Germany, Canada, and Poland, with approximately one-fourth of their publications written by a single author. Compare this to China, Spain, South Korea, and Brazil, where more than 90 percent of publications have multiple authors. The United Kingdom, which has a rate of coauthorship close to the sample mean, has the highest average number of authors per paper and is the only country among the top 15 to average more than four authors per publication.

In terms of multicountry collaboration, again, publications in which the corresponding author is European have a greater tendency to have coauthors from different countries. The western European countries in the sample (United Kingdom, Germany, Italy, France, Spain, and the Netherlands) occupy all the top spots in terms of international collaboration, whereas China, India, South Korea, and the United States are the least likely to copublish with authors from other countries.

We suggest that several features of the research culture in Europe can at least in part explain this difference in impact and collaboration between Europe and Asia. First, in the United Kingdom the main research funding bodies have invested considerable resources in getting various e-science and e–social science initiatives off the ground, particularly in the middle of the first decade of the twenty-first century. Along with these funding initiatives came regular opportunities to build community, often in the form of various "All Hands" meetings held during these same years. These regular opportunities to meet, to give and hear presentations, and to get to know the participants in the e-research space served not only to build what the funders hoped would become critical mass but also made researchers aware of each other's work, which most likely led to increased citing of publications. The likelihood of citation in this data set increases when coauthors from multiple countries collaborate on a publication: whereas almost half of single-nationality papers are cited (49.7 percent of 10,910), approximately two-thirds of papers written with coauthors in different countries are cited (60.0 percent of 1,814 two-nationality publications, 64.3 percent of 291 three-nationality publications, and 72.3 percent of 119 four- or more-nationality publications have been cited). The average number of citations per paper is also higher for papers with multiple national affiliations, increasing from 10.62 citations per paper for single-nationality publications, compared to 13.14, 12.39, and 18.48 citations on average per paper for two-, three-, and four- or more-nationality publications, respectively.

Likewise at the European level, the European Commission as part of the Sixth and Seventh Framework Programs funded work to build and implement e-research infrastructure. Given the nature of how European Commission funding operates (in which partners from multiple countries are a prerequisite for most funding calls), the collaborative projects, networks of excellence, and regular meetings arranged by European Commission bodies on topics such as e-infrastructure and research networks again have served to create awareness of related work and, as a result, to increase citation of publications.

Certain institutions are also prominent in the area of e-research (table 3.4). In the United States, Indiana University, the University of Illinois, and Purdue University form a tight triangle of e-research activity in the American Midwest. In the United Kingdom, the University of Manchester, the University of Oxford, and University College London are likewise prominent and frequent collaborators in e-research. All of the top 25 universities in e-research are, perhaps unsurprisingly, prominent institutions with global reputations for research.

Finally, we can also look at the journals in which articles discussing computational approaches to research are being published. Figure 3.5 shows the journals from this sample overlaid on a standard map of science.[16] Dark spheres represent areas of publishing activity, with the most frequent journals in the largest type and spheres.

Several things are noteworthy in this overlay. First, compared to some specialist topics, which would be limited to a small area of this map, e-research spreads across the entire map. Although some areas are more densely packed (with higher activity) and some less dense (with lower activity), some publication activity is occurring on nearly every point of the entire map. This suggests that the topics of concern here (e-research,

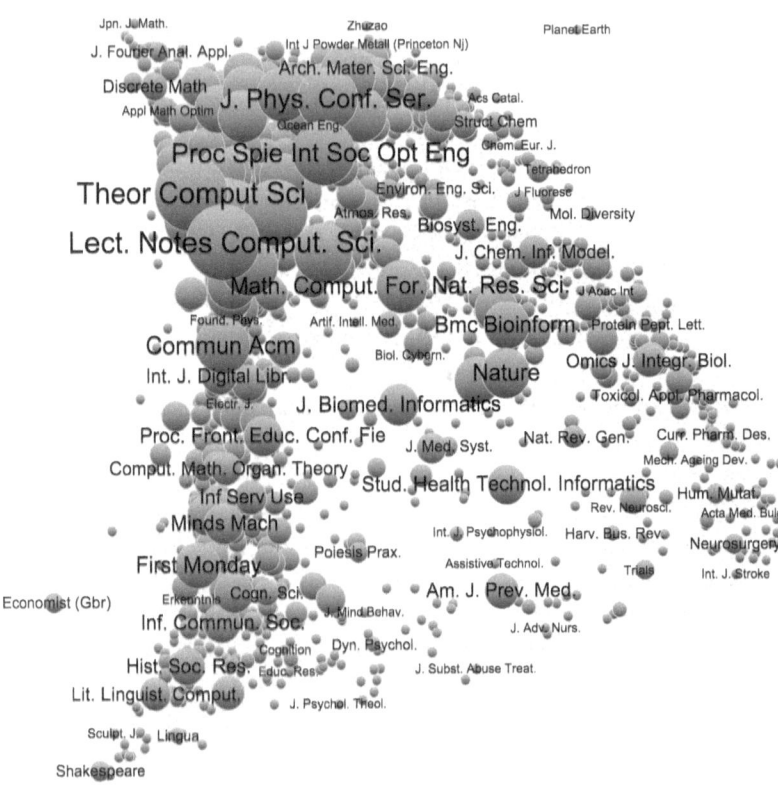

Figure 3.5
Overlay map of e-research publications.
Sources: Map generated using Leydesdorff, Rafols, and Chen (2013) and Leydesdorff, Moya-Anegón, and Guerrero-Bote (2014) methods as well as VOSViewer. Data compiled from Scopus.

e-science, cyberinfrastructure, digital humanities, etc.) are having an influence across many knowledge domains. Second, we see that there are particularly heavy areas of activity in the top left portion of the map, where computer science, engineering, mathematics, and related publications are clustered, and in the bottom left portion of the map, where the social sciences, business, information science, and communications are clustered, although this activity is most heavy in particular specialist journals such as *Information, Communication, and Society* and *Literary and Linguistic Computing*.

Taken as a whole, then, the data in this chapter tell a story: over the past 10 years, considerable momentum in the area of e-research has been growing steadily in terms of publications using the label. This growth has been led by certain countries (notably the United States and the United Kingdom), which have had the most influence in terms of numbers of both publications (bar China) and especially the impact of those publications on other scholars. Certain institutions and individuals are most prominent, but many others are engaging both in the dialogue about computational approaches to research (which these data capture) and in the day-to-day research that comes from implementing these research technologies, often without explicitly mentioning the fact (which, again, could not be captured in these data).

What we can see, then, is the messy reality of changing contours of knowledge rather than a clear-cut paradigm shift that has been advertised in the visions of e-research (such as the Berman and Brady [2005] report). Yet a different interpretation of these data can be offered: researchers are caught in a larger—albeit complex or multiform—sea change in how their disciplines (and relations with other disciplines) are being transformed, a larger change that researchers themselves may not be aware of even if they are aware of changes that affect them directly. As for e-research technologies crossing disciplinary boundaries, the picture also points to this change, but not in a uniform way: we can observe that there is an overlap between fields such as computer science and other sciences in publications, but as we will see in chapters 6 and 7 on disciplinarity, we also know that the content of the publications in both fields has different foci.

These data about "output" must be seen against the background of how researchers increasingly use online resources and are changing their work practices. The way researchers seek resources and access data sets has become a key factor in the types of information that are used (Meho and Tibbo 2003; Fry, Virkar, and Schroeder 2008). Further, as discussed in chapter 1, e-research is particularly prone to a feedback loop between increased online visibility and how researchers orient themselves toward this visibility. One crucial element is missing here: How do those who have little interest in the use of computational tools perceive the increase in e-research publications and visibility? Perhaps it is too early to say, but based on the various sources of data we have brought to bear on the topic, it can be said that there are two competing trends, at least among social scientists, who were queried in the NCeSS and AVROSS surveys mentioned at the beginning of this chapter: each suggests a picture of uncertainty in the attitudes among social scientists, with many social scientists in particular acting as "spectators" and demonstrating a lack of uptake (though it needs to be mentioned that both studies were undertaken some years ago). However, our data set shows that the number of journal publications in relation to e-research is considerable. Finally, although cross-disciplinary collaboration is mainly within computer science, this collaboration is on a sizable scale and appears to be growing.

This picture of the impact of e-research prompts some broader reflections about what effect new technologies in research have had. Michael Fischer, Stephen Lyon, and David Zeitlyn (2008) have argued that the impact of new technologies in research cannot truly be discerned until they have been in use for 10 to 20 years. In this sense, this book is tackling an emergent phenomenon. There are different time scales for what we have presented: funding, especially for infrastructures, creates a momentum for technological systems, but whether these systems can sustain themselves and what their eventual impact may be is too early to say. Citation patterns too, including how they indicate cross-disciplinary and cross-national fertilization, indicate changes that may subsequently turn out to be temporary. It is not possible to say definitively how important the gains that have been made with these tools will turn out to be, apart from noting—as we have done—the various indicators that e-research has made inroads even if it has not become mainstream.

Yet these points equally apply to the impact of any technology in the making or to new approaches to research that struggle to establish themselves against well-entrenched approaches: once these new technologies and approaches have themselves become well entrenched in 10 to 20 years, it may be difficult to discern their impact. It is also worth mentioning that the main obstacle to e-research is not active resistance to the use of new tools, but rather the inertia of using existing tools and methods. Thus, the argument that impact takes 10 to 20 years to discern can be accepted while at the same time recognizing that a partial but also unmistakably significant shift, such as the one for which we have provided various indicators, provides compelling evidence that a challenge to existing approaches, in the manner of CMs or SIMs, is beginning to make its mark. Finally, it can be mentioned that what we have gauged here are the resources that are promoting e-research and the resulting outputs: these resources are the enablers and constraints of knowledge production and how research is communicated. They are unevenly distributed, whereas the research fronts of scientific knowledge discussed in chapter 2 are global and universal. We pursue this tension in understanding science in the making, in addition to the unevenness of technology in the making—that these processes are at once ongoing and at the same time represent the state of the art for the research community at large—in the chapters to come.

4 Aggregating People and Machines: Collaborative Computation

One of the central features of computational approaches to research is the increasing dependence on the Internet not just to enable communication but also to enable the sharing of data and tools. For those old enough to remember sharing data and files via floppy disks, it is impossible to overstate the impact that networking technology has had on the practices of research. What is less well documented (with notable exceptions—see, for instance, Olson, Zimmerman, and Bos 2008; Jankowski 2009; Dutton and Jeffreys 2010b), apart from the technological changes, are the changes in collaborative research practices that accelerated in the early 2000s, due in part to the sustained and focused streams of funding discussed in chapters 2 and 3. In order to grasp the nature of these collaborative practices themselves, however, it will be useful to start with a simple two-axis diagram, which we refer to elsewhere as the "complexity continuum" (Bulger, Meyer, de la Flor, et al. 2011; Collins, Bulger, and Meyer 2011; Meyer, Bulger, Kyriakidou-Zacharoudiou et al. 2012). The horizontal axis in the complexity continuum (figure 4.1) represents computational complexity, and the vertical axis represents human collaborative complexity, and it was developed based on our research with humanities scholars and physical scientists.

The complexity continuum is a simplified way to think about how increasing computational complexity and the increasing complexity of collaborative human configurations are related to one another. Note that in the diagram, the axes are labeled "higher" and "lower" (as opposed to "high" and "low") for a reason: what is considered complex (both in terms of technology and in terms of human arrangements) is a moving window. Twenty years ago electronic full-text journals were considered complex computer-based information resources in comparison to the paper journals that researchers were familiar with; today, we all have come to expect electronic full-text journal articles as the norm. Paper-only journal articles

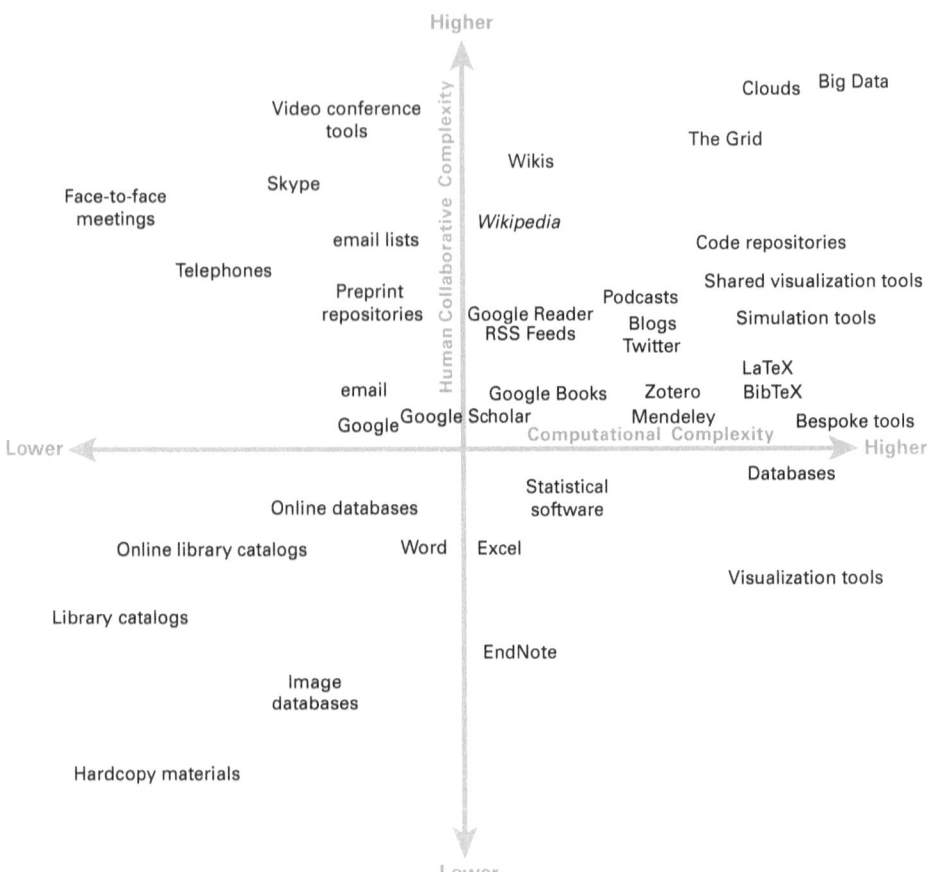

Figure 4.1
Complexity continuum.

are rare these days, although knowing exactly how many journals are still primarily paper based is difficult to ascertain. We do know, however, that reading articles from a print journal is increasingly rare, representing less than 10 percent of article reading, according to a recent study of UK academic staff (Tenopir and Volentine 2012, 46), so it would be difficult to argue that reading electronic articles is perceived to be a complex sociotechnical practice these days.

Likewise, even though today's tools and social configurations that fall into the continuum's upper-right quadrant are generally considered complex, it would be naive to think that these same tools and arrangements will still be considered complex twenty years hence. Furthermore, the *perception* of how complex a given sociotechnical configuration is depends on one's point of view: different fields, disciplines, and domains can easily perceive the same phenomenon to be either complex or simple depending on their training, biases, history, and experiences. So, for instance, whereas using the Grid is beyond many researchers' technical capabilities, for physicists working with data from CERN the Grid has become a flexible part of a daily routine (Zheng, Venters, and Cornford 2011).

The vertical axis also reflects variable complexity, but among human actors. Existing research has shown that the social complexity of collaborative relationships is often one of the biggest challenges faced by those implementing information systems and working across organizations (Orlikowski 2000; Cummings and Kiesler 2005). Nevertheless, it has been shown that research across the disciplines is growing increasingly collaborative (Wuchty, Jones, and Uzzi 2007), although the level of collaboration varies considerably depending on the field—physics compared to history, for example. It is important not just to assume, however, that physicists and computer scientists are years ahead of social scientists or humanities scholars across the board with respect to collaborative research. In a series of recent case studies of information use, we found that although on average more physical scientists were engaging research technologies (as shown in the upper-right quadrant of the complexity continuum) compared to humanities scholars, there was nevertheless considerable overlap between the humanities and physical sciences (Meyer, Bulger, Kyriakidou-Zacharoudiou, et al. 2012, 79–81).

In particular, the top-left quadrant of figure 4.1, consisting largely of communication technologies, which are relatively simple computationally yet require social and organizational coordination, is extremely important across all fields and disciplines. Email is, as it has been for several decades, still a "killer app" for researchers. Whether this remains true as new technologies develop and new generations move into research positions over the coming decades remains to be seen. Yet even if the specific applications change, the need to use simple technology to communicate quickly, easily, and cheaply will not diminish. Thus, technologies that enable writing to, speaking to, seeing, and hearing each other with relative simplicity are bound to remain important. Finally, these scales do not suggest a value judgment: tools, data, and information resources with lower complexity can easily be just as or more valuable to the expansion of knowledge as their more complex cousins. However, from a sociotechnical perspective, this relative complexity has important implications with regard to how researchers and teams of researchers engage with research technologies.

In this chapter, we argue that although there are different configurations of people and machines, the possibilities for collaborative research are not infinite. In fact, the opposite is true: the various configurations can be broken down into a few major ones that have different levels of complexity. This kind of analysis is therefore useful for understanding some of the main ways that people and machines work together to advance knowledge.

Aggregating Machines: SwissBioGrid

As we argued in chapter 2, e-research is about the development of more powerful research technologies for knowledge production, and in chapter 3 we documented how one of the goals of the e-infrastructure efforts in Europe and the cyberinfrastructure program in the United States has been to join up computing resources.[1] This goal has been implemented using a variety of technologies, including the Grid, middleware, clouds, and others. Although the technical details vary, all of these technologies share the goal of connecting computers so that their power (in terms of computing or storage or both) can be shared among research teams and facilities. To understand the technical and social complexities involved, we can take the case of a Grid-based approach to aggregating machines for life sciences research in Switzerland (mentioned briefly in chapter 1 and reported in detail in den Besten, Thomas, and Schroeder 2009).

Life sciences provide a good illustration of how research is increasingly becoming an information science that deals with large volumes of information and data. One indication is the emergence of a subdiscipline specifically devoted to dealing with the flows of data and information that underpin modern life science research: bioinformatics. Another indication is that because laboratory experiments in the life sciences yield ever increasing amounts of data, more and more time from and effort by scientific teams are being used not in wet laboratories, but in the workplaces of data analysts.

Unlike in the United Kingdom and the United States, in Switzerland there were, at the time of the research considered here, no attempts at creating a national e-infrastructure. Even without the lure of such program funding, however, in 2004 a group of researchers in Switzerland established the project SwissBioGrid to explore whether Grid computing technologies could be successfully deployed within the life sciences. SwissBioGrid was established to provide computational support for two pilot projects: one for proteomics data analysis and the other for high-throughput molecular docking ("virtual screening") to find new drugs for neglected diseases generally, but more specifically dengue fever in the original project. The proteomics project was an example of a large-scale data-management problem, applying many different analysis algorithms to terabyte-size data sets from mass spectrometry, involving comparisons with many different reference databases; the virtual screening project was more a purely computational problem, modeling the interactions of millions of small molecules with a limited number of dengue virus protein targets. As discussed more fully in this chapter, data-intensive discovery gained in this case from close collaboration with industry and harnessing distributed computing power. At the same time, the project was not sustained via a more generic infrastructure, and it is also unclear how the life sciences, which are typically organized into separate labs, are suited to such a larger system. In any event, without this integration, the benefits of this successful project could not be sustained.

The SwissBioGrid project began in early 2004 when two researchers from the Biozentrum at the University of Basel and the Swiss Institute of Bioinformatics joined together with researchers from the Novartis Institute for Biological Research (a pharmaceutical company), the Zurich Functional Genomics Center (a joint venture between the University of Zurich and the Federal Institute of Technology in Zurich), and the Swiss National Supercomputing Centre (CSCS) in Lugano (Podvinec, Maffioletti, Kunszt, et al. 2006). Thus, there were six institutions involved, but a small core of researchers, technical staff, and project managers.

In 2008, after SwissBioGrid had ended in 2007, there was a new national initiative in Switzerland, the Swiss National Grid Association, funded cooperatively by a number of Swiss universities and aiming to capitalize on the SwissBioGrid and other Swiss initiatives. Around the same time, in 2007, the Swiss National Science Foundation started funding an initiative in systems biology called SystemsX,[2] in which some of the SwissBioGrid participants have also taken part; this collaboration entered its second phase of funding in 2013 and is scheduled to continue at least through 2016. The Swiss National Science Foundation has historically been unwilling to fund infrastructure projects per se (Bairoch 2000). The fact that the new SystemsX project has explicit, although limited, funding for development of collaborative research platforms may be indicative of a greater appreciation for the needs of collaborative research in Switzerland. In early 2009, Peter Kunszt, the SwissBioGrid infrastructure coordinator, became project manager of the Swiss Systems Biology Initiative (SystemsX.ch), giving some continuity of experience from SwissBioGrid to the new organization supporting the bioinformatics community in Switzerland.

Here, however, we focus on SwissBioGrid in the period between 2004 and 2007, before the Swiss national infrastructure efforts had gotten under way, when the project had to rely on local funding and local resources only. Having only local funding and resources has some negative effects, as Michael Podvinec (of the Biozentrum and Swiss Institute of Bioinformatics) pointed out: "This grassroots way of operating has also had a lot of problems. People are going to be motivated by their own interests foremost, and only contribute when you really force them to. This will lead to broken promises, to delays and to parts of the common infrastructure that never get properly done" (interviewed by Ralph Schroeder, May 4, 2007, Basel, Switzerland).

These challenges slowed but did not prevent the project from developing workable tools. In early 2006, Arthur Thomas (one of the authors of the article on which this section is based) became the project coordinator. Soon thereafter the collaboration was expanded to include the Proteome Informatics Group of the Swiss Institute of Bioinformatics at the University of Geneva and the Institute of Molecular Systems Biology at the Federal Institute of Technology Zurich, which were also looking for computing resources large enough to tackle the enormous (terabyte-scale) amounts of data beginning to emerge from high-throughput proteomics studies, which needed to be analyzed using parallel application of different algorithms. GeneBio SA, a Geneva-based company that produces a widely used commercial proteomics analysis suite called Phenyx, provided substantial technical support and favorable licensing terms to make Phenyx accessible to SwissBioGrid. Schrödinger LLC, a commercial company that develops the widely used Glide package for modeling ligand drug/protein binding, also cooperated to make its software available on favorable licensing terms. And so SwissBioGrid served as a sandbox in which a multitude of players could experiment with new technologies for distributed computing. The collaboration paid off scientifically as the combined effort of about two dedicated full-time-equivalent researchers per year resulted in a total of 32 publications in the three years that SwissBioGrid operated.

From a technical perspective, SwissBioGrid's main goal was to develop an infrastructure that could use the distributed-computing capacity of clusters and PCs at the partner institutions to solve hard computational problems in biology. This capacity included Unix clusters at Vital-IT, CSCS, the Biozentrum, the University of Zurich, and the Federal Institute of

Technology Zurich as well as PC farms at the University of Basel, Novartis, and the Friedrich Miescher Institute. This infrastructure served two scientific goals. First, it served as an environment for high-throughput molecular docking ("virtual screening") to find new drugs for neglected diseases (specifically, again, for dengue fever, which is widespread in mosquito-ridden areas of the world), and later it served as environment for proteomics data analysis as well. The proteomics project was an example of a data-management problem, applying many different analysis algorithms to terabyte-size data sets from mass spectrometry, involving comparisons with many different reference databases; the virtual screening project was more a purely computational problem, modeling the interactions of millions of small molecules with a limited number of protein targets on the coat of the dengue virus.

It is important, however, to distinguish between the computing environment and the research environment. Crucial for the scientific success of the virtual-screening project was that Novartis, the partner from the pharmaceutical industry, agreed that its Tropical Diseases Research Institute in Singapore would perform experimental studies on any leads that emerged from the virtual-screening experiments. Novartis further committed itself to provide any drugs that resulted from this collaboration at cost to dengue sufferers in the developing world. This project was thus a good example of a public–private partnership for the public good. According to Michael Podvinec and Torsten Schwede (another researcher on the project), many bioinformatics studies deliver DVDs full of data that are never looked at. As Schwede put it, "There have been a number of projects that have done, or have tried to do computational drug discovery ... but which end up with saying 'We have been very successful; we produced 30 DVDs full of results.' And that's as far as it goes. It usually has a footnote: 'these are now very valuable results for biologists who will test this' ... or not" (interviewed by Ralph Schroeder, May 4, 2007, Basel, Switzerland; ellipses indicate deleted material).

The collaboration with Novartis made sure that the results discovered using *in silico* computational methods were validated. In return, Novartis was given the exclusive right to exploit the results. Because the dengue virus was of limited commercial interest to Novartis, establishing a public–private partnership with it was relatively easy. Even so, the legal negotiations for this part of the project were among the most difficult aspects of the whole project (with respect of the amount of time they took).

In the course of the project, significant software-development efforts were undertaken by the Biozentrum, where Podvinec developed a mechanism for dispatching jobs to a mixed grid of desktop PCs and Linux clusters, and by CSCS, which managed the overall Grid infrastructure. For some key technologies, the best solutions were available only from commercial vendors. Unfortunately, most existing licensing models turned out to be incompatible with Grid computing because they relied on technical assumptions not necessarily satisfied in a Grid environment (such as continuous network connectivity to a license server) or on pricing schemes originating from the pre-Grid era (e.g., one license per central processing unit). Luckily, at least the two companies already mentioned, GeneBio and Schrödinger, were prepared to provide their software on favorable terms. More cumbersome than the licenses, however, were the differences among software and database interfaces that had to be bridged, the numerical stability of software in heterogeneous environments, and the problem of data caching.

The incompatibility of databases is the result of how each one came into being: many small groups of researchers work on different and separate databases, which use different languages to describe their content (or even in some cases the same language to describe different content!). Similarly, software in computational biology and bioinformatics has often been developed on a single computational platform without much emphasis on portability or scalability: here we can see limits, which are perhaps more evident in the life sciences than elsewhere, to the transferability that we argued in chapter 2, following Shinn and Joerges (2002), is characteristic of research technologies. Many software packages used in this field give different numerical results when executed on different hardware or software platforms. These differences are often *not* negligible and might lead to contradictory scientific conclusions depending on the platform where a calculation is performed. Thus, in the case of SwissBioGrid, validation of numerical stability turned out to be essential for using heterogeneous Grid architectures. A final obstacle was the need to ensure that up-to-date, synchronized copies of the needed databases were available "on demand" at the computational nodes. Their large size precluded either keeping copies everywhere or moving copies to the computational nodes "just in time." For the time being, the problem of optimizing database distribution (sometimes called the "data caching" problem) remains without a satisfactory generic solution. And so SwissBioGrid had to develop specific solutions to that problem as well.

From Schwede and Podvinec's perspective in particular, the technical expertise developed through SwissBioGrid represented a somewhat wasted

Aggregating People and Machines 77

effort in two ways. First of all, they were unlikely to receive much scientific recognition for all the efforts they put into making the tools work before the tools started to produce results. This is a typical problem in e-science or cyberinfrastructure, where there is little reward for tool development as opposed to high-ranking publications within the researcher's own discipline. Second, the extent of adaptations that needed to be made and the brittleness of the system that resulted meant that their continued expertise would be required if they wanted to make their tools work for other users or in other environments. It could be worse, Schwede acknowledged: "At this stage we are very lucky because it is kind of pioneering work. But let's imagine we would have been successful in providing a service structure: running an infrastructure service is much more work than building a prototype. And if you run a service that is acting in the background—running as middleware—you don't get any recognition for this at all." Yet Schwede and Podvinec's motivation to engage with the Grid in the first place was the hope that it would provide a more user-friendly environment that others would then maintain.

At the start of the project, researchers hoped to take advantage of "middleware"—that is, system software for the management of distributed resources, which was being developed under the auspices of the European Union–sponsored EGEE project (mentioned in chapter 1; see also Schroeder 2008) for the analysis of observations from CERN's LHC. However, in this event, the NorduGrid/ARC suite was chosen as middleware because the more widely used EGEE/gLite suite, after a comprehensive evaluation performed by CSCS and Vital-IT, appeared to be too complex, inflexible, and intrusive, especially when desktop computers were also used as computing resources. In addition, because biological problems tend to involve more complex and heterogeneous data than many in the physical sciences for which both NorduGrid/ARC and EGEE/gLite had been developed, the project had to overcome substantial challenges in distributed-data management, such as the "data-caching" problem mentioned earlier, before being able to deploy the middleware. As Podvinec explained,

I think that the reason why we started with the dengue project is because it is so close to High Energy Physics that in the end you can see it as another incarnation of this needle-in-the haystack problem with just a little bit of bio-typical problematic in it. ... But if you take the next step, as in the proteomics project, you already need the context of all known genomes for a search, and you need to distribute these large datasets to all computers running the computation. (interview; ellipses indicate deleted material)

In this context, NorduGrid/ARC had the additional benefit that its developer community was relatively small and hence more open to demands for features that were necessary to make it work for SwissBioGrid. Yet in the long run this benefit proved to be a disadvantage because uncertainty about funding for continued development of NorduGrid/ARC made it unlikely that the tools that were so well integrated with this suite would remain easy to deploy in future.

Data-driven discovery, according to the SwissBioGrid experience, is a collaborative effort, which includes many partners from industry as well as from academia. In fact, the scale and capital requirements of modern laboratory research often make ties with industry indispensable. Much of the expertise that is accumulated in the process of discovery relates to tool development. As such, the scientific value of this expertise is not always immediately obvious because it relates to matters that are often at best peripheral to biologists' concerns. And this low value, in turn, makes it hard to obtain credit for tool development, as opposed to the credit given for the subsequently derived scientific results. Moreover, tool development is not limited to the outer layer of end-user applications: SwissBioGrid found that it is not always safe to assume that software infrastructure for distributed computing that has been developed for data-driven discovery in one field—say, particle physics—can be adopted to discovery in another field—say, life sciences—without substantial adjustments.

The heterogeneity of resources in the life sciences and the recurring need to combine multiple resources are major factors that make the establishment of a single infrastructure difficult. As Graham Cameron of the European Bioinformatics Institute explained,

The behavior of the molecules themselves is very complicated, and the kind of information that we can collect on it is very complicated; there are actually a lots of bits of biology that sort of talk to each other—there are always the DNA sequencing guys; there are the protein guys; there are the molecular interaction guys; there are the pathway guys—they're all talking about a common set of entities in a living system; and they are collecting what is currently rather patchy information, scientific information, about those; and you want, both within any individual collection, if you're talking about a given protein product, you want to call it the same thing, the connections between all those entities come from the identifiers and the knowledge attached. ... There are a lot of places where it is relatively easy, [others where] you'd discover that biology did something that you didn't anticipate. So, the structure is constantly evolving; the entities that are in there are connected to each other; you have to share the data-structures, and you have to share the way you refer to the entities. And the entities don't stand still either. You know, it is commonplace for people to be blathering on about some gene or other and a couple of variants of it only to discover that actually it wasn't one gene that they were talking about, but two different genes. (interviewed by Ralph Schroeder, May 21, 2007, Basel, Switzerland; ellipses indicate deleted material)

The life sciences make use of a wide variety of databases (meant in a rather loose, nontechnical sense) in a wide variety of formats and, more importantly, with a wide range of semantics. So integrating information from different databases is made more difficult by the difficulties of reconciling those semantic differences. New tools, largely coming from the "ontological engineering" community, hold out the promise of being able to identify inconsistencies of nomenclature between different databases and perhaps even to reconcile those differences. Significant effort (exemplified by the Open Biological Ontologies project[3]) has been put into addressing this problem. As Cameron explained, "We need to connect the databases with each other so that ... not only can people find things of interest, but they can build tools, which are satisfactory in terms of the visualization of that information" (interview).

In SwissBioGrid, the proteomics project in particular was challenged by the need to compare the experimental data with reference (peptide fragment) data from a number of different sources and to merge the results of different analysis methods to create a coherent and integrated view of the results.

Moreover, the problem of establishing appropriate reward structures remains unsolved. Many academic institutions are wrestling with this problem internally. Developments in wiki-based collaborative knowledge creation (such as WikiGenes, WikiProteins) make possible the explicit tracking and visualization of individual contributions to the knowledge repositories, yet how this information can be applied to traditional methods of impact analysis remains to be seen.

As a consequence, the organizational structure that typifies research in SwissBioGrid is difficult to pin down as well. The laboratory no longer seems to occupy the dominant place that it occupied several decades ago (for examples of this change, see Power 2011). Yet reduced dependence vis-à-vis the principal investigator of a single laboratory (as described in Knorr Cetina 1999) seems to have been replaced by new dependencies due to the need to retain access to data, to ensure the validation of results, and to sustain research procedures in changing environments. Alternatively, if the visions of cyberinfrastructure, e-science, and the Swiss National Grid Association materialize, the life sciences might become part of nationally and internationally organized technologically integrated bureaucracies.

Yet, if anything, the SwissBioGrid experience highlights the difficulty of technological integration without conceptual integration and without a sustaining infrastructure.

Hence, from a research policy perspective one might wonder whether it would not be better for the life sciences if their research organization could be transformed into the conceptually integrated bureaucracy that is more typical of particle physics or into the fragmented adhocracy that characterizes parts of the social sciences, where data sharing is developing: a bureaucracy might be more efficient and thus allow for the creation of a central information or data infrastructure. Yet an adhocracy might be more diverse and thus require a more flexible infrastructure or simple federated resources. The life sciences could become more like a conceptually integrated bureaucracy if relatively big institutions, such as the US National Center for Biotechnology Information, the European Bioinformatics Institute, and the Wellcome Trust's Sanger Institute, were able to secure more substantial long-term funding to tackle infrastructure issues, as opposed to purely scientific ones. These institutions would then attract the necessary support staff to develop the custom infrastructures needed to carry out this research, much like CERN in the context of particle physics.

Alternatively, the life sciences could become more like an adhocracy if more efforts were made to guarantee easy and equal access to data, publications, and tools (as we see in the case of GAIN, chapters 1 and 5). The newer paradigm of "Cloud computing" might also provide computing to researchers at large where the Grid paradigm appears to have fallen short, or the wrapping of research tools and data as web services might make new configurations for research easier to achieve. In addition, substantial investment in traditional laboratory facilities would be required to promote the *in-vitro* validation of interesting *in silico* research. These questions of resources and organizational structures to promote and sustain sociotechnical systems for research clearly also require momentum from CMs, as described in chapter 2.

Perhaps it is not necessary or even possible to push the life sciences into the mold of an integrated bureaucracy. Steven Vallas and Daniel Kleinman (2008) characterize the life sciences as a "heterarchy"—an increasingly interconnected scientific field in which the distinctions between academic and corporate sciences are blurred and that is marked by the anomalies, tensions, and ironies already observed by Edward Hackett (2005). For research that is perhaps located in what Donald Stokes (1997) calls Pasteur's quadrant of use-inspired basic research as opposed to "pure" basic research, it is hard to see how a neat separation between academic and commercial interests will be maintained. The questions are, rather, If the academic and corporate spheres are no longer separate, what organizational principles should govern the common sphere that emerges, and how should potentially contentious issues such as intellectual property ownership be settled? Vallas and Kleinman (2008) argue that the convergence to date has been asymmetrical in that universities have been more eager to adopt the behavior of industry than vice versa. There are also signs, however, that industry is adopting an academic style of research as well as focusing on open innovation in basic research such that it can reap more benefits from the results (Hope 2008).

In this section, we have looked at the organization of research in one particular project, SwissBioGrid. What we found is a project that was the product of a collaboration that spanned multiple places and disciplines and relied on the synergy of corporate and academic research. It also required a complex and challenging level of technical integration. In this case, it is worth mentioning the public–private partnership between the Biozentrum, Schrödinger LLC, and the Novartis Institute for Tropical Diseases was crucial to SwissBioGrid in terms of distributed computing. This type of industry–university interaction is often seen as a major challenge to new ways of doing research (it has been explored, for example, by Joe Fore, Timothy Lenoir, and their colleagues (Fore, Wiechers, and Cook-Deegan 2006; Lenoir and Giannella 2006). Yet we found that this interaction did not present a particular problem in the case of SwissBioGrid, though the legal complexities were a delaying factor at the start.

The challenge came instead from the technical and organizational issues of data management and computational problems in the life sciences. We highlighted the successes in tackling these problems but also noted the heterogeneity of the life sciences and the lack of an infrastructure in which the solutions developed in SwissBioGrid could be sustained. What we also see in this case are the difficulties in providing an account of a (research) technology in the making: SwissBioGrid as a project was discontinued even though it had successes in its technical aims. The infrastructure that could have sustained the project has subsequently begun to emerge, and the expertise and technologies that were part of the project have become part of these new efforts. Analysis of knowledge production captures an ongoing process, whose many component parts are constantly congealing, being built upon and extended, and moving in new directions.

Aggregating Many Eyes for Science: Galaxy Zoo

Although the Grid and other distributed computation models—such as the Grid's immediate predecessor, Beowulf Clusters, or its immediate successor,

Cloud computing—require human interactions for their implementation and uses, as we have seen, these configurations are at their core primarily technological constructs around which, as argued in chapter 2, researchers' efforts to produce knowledge can focus.[4] Technical considerations were among the most important driving factors in the design of SwissBioGrid, and they crystallized in a sociotechnical system. The Grid's affordances, processing on a large scale, and their relative strengths and limitations for the life sciences in general and SwissBioGrid in particular, were at the center of the choices being made.

Not all scientific problems, of course, are amenable to algorithmic approaches as implemented by computers. Even as computers become more powerful and flexible, for some tasks the human brain remains far more able to perform the calculations necessary to solve problems. One area where this is true is in certain types of pattern matching. Although computers are able very successfully to match certain kinds of patterns against large databases more quickly than humans can, the human brain performs better for other types of patterns with ambiguous shapes and outlines. We come across another example of this later when we discuss how marine mammal scientists identify whales and dolphins, but our example here is even less close to home than the oceans: identifying galaxies.

The goal of Galaxy Zoo—which we briefly mentioned in chapter 1—was originally to classify galaxies from images provided by the Sloan Digital Sky Survey (SDSS), although the project has expanded considerably since then. The first Galaxy Zoo project was a response to a data-deluge problem (Hey and Trefethen 2003) experienced by astronomers: the SDSS produced far more photographs of the distant sky than previous projects had done. The initial Galaxy Zoo sample had almost 900,000 objects, whereas previous work with SDSS data ranged from 2,500 to about 50,000 manually classified objects (Lintott, Schawinski, Slosar, et al. 2008). Astronomers could not inspect the entire catalog, especially since multiple independent classifications of each galaxy are needed if researchers are to have confidence in the results (Lintott, Schawinski, Slosar, et al. 2008). Part of the survey had been professionally categorized before the Galaxy Zoo project began; this categorization provided a baseline against which to measure the citizen-science contributions. The main tasks for scientists working on the project then became to encourage citizen scientists to classify data to contribute to scientific findings. The scientists' roles were thus extended from routine practices in astrophysics because they had to interact with the general public via the online platform they built for the project,

Zooniverse. In addition, the project required several full-time software engineers who would work on developing and improving the Zooniverse cyberinfrastructure. Scientists working with the public on Zooniverse projects are known to them as "Zookeepers," and the public call themselves "Zooites" (Raddick, Bracey, Gay, et al. 2010).

Citizen scientist contributors register with Zooniverse and then choose which project they want to contribute to. For those participating in Galaxy Zoo, an image of a galaxy is shown in the browser, and the user clicks one of six buttons on the right of the image to classify the type of galaxy (Raddick, Bracey, Gay, et al. 2010). For more complex projects, the user may be asked to identify more features or types of object or to measure objects by selecting them.

The Zooniverse team was surprised by the number of contributions to the first Galaxy Zoo, although the strong public response was aided by mainstream media publicity. The original Galaxy Zoo project launched on July 8, 2007, and was covered by the BBC on its website and a morning radio show on July 11, which was followed by coverage by other news outlets. Within one day of this coverage, nearly 1.5 million classifications had been completed by more than 35,000 volunteer classifiers (Raddick, Bracey, Gay, et al. 2010, 3). Over the two years of the first Galaxy Zoo project, around 70 million classifications were made by more than 180,000 volunteers.[5] The scientists working on the Galaxy Zoo project estimate that the data provided by the citizen scientists is equivalent to the project maintaining approximately 150 full-time classifiers over five years.

In Galaxy Zoo, each astronomical object was originally classified approximately 40 times by different citizen scientists to provide confidence in the classifications, although later techniques have allowed the research team to reduce this number. Such confidence is reinforced by comparing the individual citizen scientists' community classifications with professionally qualified objects. This comparison also allows the scientists to rate the classification quality of individual citizen scientists. These data comparisons mean that obviously wrong answers can be quickly discarded and, as Zooniverse's technical lead remarked, "Internet trolls get bored pretty quickly with us" (interviewed by project staff, July 14, 2011, Oxford, UK) because they are not able to reliably disrupt the classifications.

The scientists consistently refer to the Zooniverse projects as a collaboration between scientists and citizen scientists. Thus, there are two facets to the collaborative work in the Zooniverse: first, the collaboration between the scientific sites involved in the collaboration (Oxford University, University of Nottingham, University of Portsmouth, Yale University, and

Johns Hopkins University) and, second, interaction and communication with the citizen scientists.

Interaction with the citizen scientists is mainly via the Zooniverse forums and the scientists' blogs on the Zooniverse website. Although the majority of citizen scientists do not use the forums, those who do are generally very active, and the community that has grown up around the forums has been central to making several new astronomical discoveries. The Zookeepers are careful to choose new projects that allow the citizen scientists to contribute scientific data: "I think it's really about giving them credit," stated Chris Lintott, "and it's not us against them, or science team versus them doing work for us. It's really sort of a very collaborative effort" (interviewed by project staff, July 19, 2011, Oxford).

Questions of credit and acknowledgment in authored papers are still being developed and clarified as the project progresses. In general, the scientists are the authors of the papers, and the citizen scientists who have made significant contributions to the papers are recognized in the acknowledgments. However, the scientists also consider some citizen scientists' observations worthy of an authorial credit, such as Dutch schoolteacher Hanny van Arkel's discovery of a new astronomical object, Hanny's Voorwerp (Lintott, Schawinski, Keel, et al. 2009).

Citizen science data has transformed, at least in this project, the visual analysis of astronomical image data. The citizen science data from the Galaxy Zoo and other projects can be used to train computer classifiers—and if computers can become as accurate as humans, it will transform the scale at which astronomical data is produced and enable new questions and further new discoveries. The Zookeepers are investigating this approach—in 2010, they published a conference paper titled "Data Mining the Galaxy Zoo Mergers," which examines the feasibility of several approaches to identifying "correlations between human-identified patterns and existing database attributes" (Baehr, Vedachalam, Borne, et al. 2010). Steven Baehr and his colleagues found small information gains but also identified promising directions for further studies in this area. Success in this approach would help to resolve a possible new data deluge in astronomy: "when the next-generation telescope comes along ... and it produces 100 billion galaxy images instead of 1 million," data-mining techniques will be essential because even 400,000 Galaxy Zoo volunteers would take many years to process such an amount of data.

Several new discoveries have been identified from this project—for example, Hanny's Voorwerp and a new type of galaxy called "Green Peas," discovered with the public's help. New Zooniverse projects are extending the project into new disciplines, which, as mentioned earlier, illustrates how research technologies "travel." For instance, the Ancient Lives papyri analysis project employs a postdoctoral researcher "jointly appointed between Physics and Classics" at the University of Oxford to use similar crowd classification methods to understand ancient papyri.

Most of the other groups of physical scientists interviewed for the study that included the Zooniverse case (Meyer, Bulger, Kyriakidou-Zacharoudiou, et al. 2012) asserted that although technology made their work much faster in terms of the amount of information they could find and process, the tools did not allow them to explore completely new areas of their fields; rather, they were doing "more of the same." The Zooniverse project is similar in this regard, except that "more of the same" has exceeded expectations, with contributions by citizen scientists increasing data analysis exponentially. The project has also allowed a rediscovery of the use of visual morphologies (i.e., classification based on visually assessed and thus labor-intensive criteria) in astronomical research. As one participant noted, "Because the sample sizes of galaxies had gotten so large, there had been a trend towards ignoring visual morphologies" and instead using factors such as color and concentration that can be measured computationally; in other words, there has been a shift in research methodology in addition to a scaling and speeding up of the work.

Astronomers have a long history of using computers to store and analyze their observations, but they have been slower to formalize computational methods as a way of extending the field. An astronomer who is also the technical lead on the Zooniverse projects hoped to train researchers in computational methods in astronomy: "I think they [biologists] realized that there was a whole area of specialism—you know, data-intensive biology. They named it, they called it bioinformatics, and it—you know, this rare computational methods applied to their research area. And what surprises me about astronomy is that astronomers haven't done the same yet. There is a term—astroinformatics—but it's not very widely used" (anonymous informant, interviewed by project staff, July 14, 2011, Oxford). Perhaps, as with bioinformatics, a specialization will develop in astroinformatics, or computational methods may simply become part of all astronomers' standard methods. Or perhaps these techniques will remain relatively less developed in astronomy when compared with their use in fields such as biology.

In terms of developing further work, the Galaxy Zoo team is refining the user interfaces for the projects. They realized that it is necessary to make better connections between the citizen scientists and scientists because "one of the main problems with having more projects is that we need people to filter the important questions that only we can answer" (Galaxy Zoo team member, interviewed by project staff, July 14, 2011, Oxford). Further ideas include a general journal discussion, similar to journal clubs in the biological sciences, via Twitter or Skype or both, "where you can sit down with a paper and go, 'I don't understand Figure 3.' Or 'I think this paper's completely crazy,' and someone else explains why they don't think it is" (Galaxy Zoo team member, interviewed by project staff, June 7, 2011, Oxford).

The Zooniverse case raises a further question about how new technology can support efforts such as engaging the public with science: the technology to enable such collaboration is fairly simple, certainly far simpler than the telescopes that gathered the data. However, the simple web-based technology harnessed a much more complex system—many human brains—that still has no parallel in computer-based tools. How these and other methods are used to enhance collaboration and the analysis of digital data in science provides one example of our conceptualization of e-research, the harnessing of many minds to an online task.

Aggregating Many Minds for the Humanities: *Pynchon Wiki*

Just as the Galaxy Zoo approaches takes a problem that (at least currently) is more readily solved by distributed human eyes than by distributed machines, many projects in the humanities have started using crowdsourcing techniques in the past decade.[6] These techniques range from asking people to help tag photographs, as in the Library of Congress Flickr Commons[7] project (Vaughan 2010), to contributing personal materials to archives, as in the Great War Archive.[8] The case we turn to next—also mentioned briefly in chapter 1—examines how relatively simple technology was used (as with Galaxy Zoo) to aggregate people to solve a new challenge: annotating an author's work. This was done with *Pynchon Wiki*, a project that sheds light on collaborative digital research in a domain that on the face of it seems alien to computational approaches: literary studies.

Ward Cunningham developed wikis in 1994 to allow a relatively quick and simple way of editing web pages (Cunningham 2012); the most well-known product of this tool is *Wikipedia* (Reagle 2010), though there are many other wiki-type projects. The wiki discussed here is devoted to the contemporary novelist Thomas Pynchon, a writer who is regarded as one of the most important postwar American authors.

When his novel *Against the Day*, weighing in at 1,085 pages, was published in November 2006, a wiki[9] sprang up immediately that allowed literary sleuths to collaborate in annotating the whole novel within a few months. A number of wiki sections have subsequently been created on the same site for his other novels. Before *Pynchon Wiki*, there was the Pynchon-l mailing list,[10] and several web pages were devoted to Pynchon's work.[11]

Pynchon is widely regarded as a difficult author. His work is filled with so many complex topics and allusions that, as needed for only a few other authors and works of fiction (such as James Joyce's *Ulysses*; see Gifford and Seidman 1974), his novels have generated secondary works that provide a reference guide and scholarly apparatus for interpretation. These annotations help readers and scholars navigate Pynchon's novels, and before the advent of the Web, several such companion volumes were produced in printed book form for Pynchon's work. Steven Weisenburger (1988) produced one such annotated edition for Pynchon's most famous novel, *Gravity's Rainbow*.[12] Online annotation, as with the *Pynchon Wiki*, is quite different, however, insofar it allows many people to collaborate on this research and allows the process to be open-ended.

It is important for understanding *Pynchon Wiki* and especially for those who are not familiar with this author to emphasize how central the range of obscure references are to his work. Pynchon's novels are encyclopedic in that he uses in-depth knowledge of many esoteric topics; indeed, he revels in arcane areas of popular culture, history, science, and many other areas. A wiki can be useful in his case because, apart from cataloging and elucidating these details, it can either corroborate Pynchon's research—to see if he has done his homework, so to speak—or ferret out if he is fabricating these details. In the latter case, in turn, there are two possibilities: one is that this fabrication is deliberate—novels are fiction, after all—but it might also be that his account or his details are wrong or implausible, which can detract from the novel and the many "local pleasures," as the critic Michael Wood (2007, 12–13) puts it, that a Pynchon novel provides. Regardless of whether the status of Pynchon's work rests in part on this esoteric research, being able to recognize and understand his sources enhances the appreciation of his fiction.

Pynchon's work very much encourages this kind of exercise of checking up on him, even if it is debatable how much hangs on the accuracy of his sources and allusions. Pynchon has always made fun of academic seriousness and of the enterprise of detective-like attempts to get to the bottom of things. And yet, although making fun of scholarship, Pynchon admits that the facts matter. In the introduction to his short-story collection *Slow Learner*, where he looks over past mistakes, he says:

> Though it may not be wrong to absolutely make up, as I still do, phony data are often deployed in places sensitive enough to make a difference, thereby losing what marginal charm they may have possessed outside of the story's context. ... The lesson here ... is just to corroborate one's data, in particular those acquired casually. ... We have, after all, recently moved into an era when, at least in principle, everybody can share an inconceivably enormous amount of information, just by stroking a few keys on a terminal. There are no longer any excuses for small stupid mistakes. (2000, 16)

These arcana are also integral to his novels' storylines. One of the main themes of *Against the Day*, as in his other novels, relates to how the heroes—or antiheroes—chase a trail of clues to try to get "behind" a string of seemingly interrelated events. These obscure topics could therefore be important: they could be clues in a detective-type novel! But this trail also misleads because, as any reader who is familiar with Pynchon's work will know—unlike readers who are new to him and unlike characters in the novels—these clues are also red herrings, and there will be no final resolution to the mystery. In other words, his novels can be seen as a wild goose chase.

As we shall see, the detective work that Pynchon's work invites, whereby many collaborators can follow up, connect, and cover a wide variety of topics from very diverse areas of knowledge, is also one of the strengths of an online wiki-based tool. A wiki can easily deal not just with a wide range of topics, but also with how they may or may not fit together—for example, by cross-linking more extensively than is possible with an index.

A final characteristic of Pynchon needs to be mentioned, which is that he, like J. D. Salinger, is legendary for his reclusiveness. Little is known about his personal life; his whereabouts are not publicly known; and he makes no public appearances. To put it sociologically, unlike other authors who can be an offline "sacred object" of worship that offers many possibilities for readers to engage with him or her (book signings, media interviews, gossip about their lives), in Pynchon's case the online realm is perhaps the only form of interaction readers can have with other fans—something on which fans can jointly focus their worshipful attention. An online forum that allows fans to do this is thus perhaps more important for Pynchon than for other authors who are available to explicate their work or provide other materials for fans.

The context for the emergence of the wiki for Pynchon's 1,085-page novel *Against the Day* (2006) was the considerable online "buzz" that had been building up on the Web in the months leading up to its publication (Patterson 2006; *Pynchon Wiki* 2007). The wiki began immediately once the novel was published, and the rapidity with which contributions were made and the number of contributions can essentially be seen as a race to the finish line: a rapid surge of activity by contributors for two or three months of hectic activity, rising to a peak early in 2007 when the pages were "complete," in the sense that entries had been made for each page in the novel, and then rapidly falling off by the summer when the entries had consolidated, at which point the activity plateaued on a low level as entries were corrected or added to.

Tim Ware, who was also responsible for a Pynchon webpage[13] and who has a background in web page design, created *Pynchon Wiki*. The policies whereby he manages *Pynchon Wiki* are rather informal. Unlike in *Wikipedia*, participants who want to edit a page have to register for *Pynchon Wiki*, and Ware personally processes several new accounts per day in a user community that had grown to more than 500 as of March 2007 (Ware, personal communication to the authors, March 6, 2007). This gatekeeping is mainly to check that no malicious users who might do damage to the site use it. Although Ware has checked back with a couple of persons whose email addresses aroused suspicion when they tried to register, he has eventually accepted all those who have registered. Ware thus sees himself as a "laid-back sheriff" (personal communication to the authors, March 6, 2007), and there has been no misbehavior to date. In terms of how he relates to the contributors, he reports that he does not know any of the major contributors offline. Regarding the level of effort involved, after the initial setup, the wiki has taken an hour or two of Ware's time every day—it is easy to see that a great deal of volunteer effort by one person is needed—and he estimates that by March 2007 he had spent more than two months or so altogether on the project.

As for *Pynchon Wiki*'s features, apart from the homepage with news, instructions for use and for contributors, and a few other functionalities and links, including articles and reviews, the main content comprises the individual entries for pages, which are organized by clusters of pages. An attempt was made more recently to cross-reference the page entries with the alphabetical index, but this process does not appear to have been completed. Weisenburger's (1988) book also provides an index of annotations but not the advantage of being able to click through from pages to entries and vice versa.

There has been some debate on the Pynchon mailing list that is similar to debates about *Wikipedia*: Is it possible to achieve high-quality annotation with so many unvetted contributors? Unlike with *Wikipedia*, however, discussions and disagreements take place on the main annotation pages themselves rather than on separate "talk" pages. The success of *Pynchon Wiki* can be gauged by the quantity and quality of the contributions it has attracted. One way to gauge the quality is to compare *Pynchon Wiki* with Weisburger's (1988) annotations of *Gravity's Rainbow*, essentially a comparison between distributed collaboration and single authorship.

In Weisenburger's book, if we count as "entries" all the items listed in the index, we can say that 904 annotations were made by one "contributor," counting only the author (or by 22 contributors, counting the number of people Weisenburger acknowledges, although he says that he drew on many more). If we estimate the word count for the Weisenburger book (using 12 words per line and 45 lines per page in a book of 300 pages, although many pages have some empty space), we arrive at approximately 162,000 words. In the case of the *Pynchon Wiki* annotations for *Against the Day*, in July 2007 there were 455,057 words and 235 contributors (see Schroeder and den Besten 2008 for how these figures were calculated). It can be seen that the *Against the Day* wiki contains roughly three times as many words for a novel of roughly similar size as *Gravity's Rainbow* (1,085 pages in *Against the Day*, 760 or 887 pages in *Gravity's Rainbow*, depending on the edition of the novel).

If one looks at the annotations in *Pynchon Wiki* in more detail (again, see Schroeder and den Besten 2008), one can see that a core of contributors is responsible for most of the edits, and one also sees that the page-by-page style of annotation has been highly successful in attracting a large number contributors to make annotations quickly as readers make their way through a book. This seems an obvious way to interpret how contributions rose quickly then declined once entries had been made for all pages and continued at a lower level thereafter as readers returned to entries to correct them and add depth.

A different way to compare the two forms of annotation is to compare sample entries from the *Against the Day* wiki with entries in Weisenburger's book. One finds that both contain errors, though the Weisenburger book tends to have a more definitive tone and contains fewer but apposite sources; *Pynchon Wiki* is more chatty and refers to more sources (Schroeder and den Besten 2008). Perhaps the main difference, however, is that *Pynchon Wiki* makes use of the latest technological capabilities, such as linking to a variety of online sources (online descriptions of places and online dictionaries), and many graphical elements (photographs, maps, and the like), an option that the book does not have.

Pynchon Wiki, like *Wikipedia*, continues to expand, but, according to Simon Rowberry (2012), the activity has flagged considerably since the initial editing period. Unlike *Wikipedia*, which brings together generally accepted knowledge, *Pynchon Wiki* contains material that is both interpretive and of interest mainly to a specialized audience. And unlike Weisenburger's book, which had to be finalized (except via the publication of an updated edition; see Weisenburger 2006) purely by virtue of its book form, the *Against the Day* wiki can at least in principle continue expanding and correcting itself indefinitely. It provides greater scope for being an endless playground of interpretation and detective work. This open-endedness has advantages and drawbacks for both the producers and the users of these wiki-based works (How much depth does a user need?). On the open-ended side, a similar question has arisen regarding *Wikipedia* over the "stabilization" of particularly contentious entries (Tatum and LaFrance 2009). In the case of *Pynchon Wiki*, this question is tied, again, to whether the wiki annotations aim to be an academic reference guide or to adopt a more playful approach that treats the text as having endless scope for further interpretive readings.

As with *Wikipedia*, authors cannot be identified on *Pynchon Wiki*; Weisenburger's single-authored book, in contrast, brings with it academic recognition. The principle of anonymity in *Pynchon Wiki* and *Wikipedia* can of course be compromised when users identify themselves on their user pages or on the discussion pages, though these pages require logins.

Thus, the main advantage of distributed collaboration in this case is speed. The wiki had covered all pages of *Against the Day* within three months of publication, whereas Weisenburger, to judge by the account in his acknowledgments (1988, vii), took several years to put together his book. And the publication process for academic books often adds at least a year to that process, whereas on the wiki publication is instantaneous once a contributor is registered. It is also interesting to see in this case that collaboration was structured as a kind of "race to the finish," which would not have occurred, for example, if the entries had been organized alphabetically or topically. Wikis thus seem to be a good tool for tasks where endless detective work is called for. *Pynchon Wiki* also shares with *Wikipedia* the feature that contributors are largely hobbyists rather than scholars and can learn from each other (see Bryant, Forte, and Bruckman 2005).

Yochai Benkler (2006) has argued that *Wikipedia*-type or "open-source" peer-to-peer collaboration can be a rich source of cultural creativity. It remains to be seen if *Pynchon Wiki* will become primarily a reference resource or an online community that continues to enjoy interpreting

Pynchon's work indefinitely. Of course, it can continue to be both if there is enough interest to sustain the latter. Apart from providing a resource for Pynchon readers, *Pynchon Wiki* thus illustrates successful online collaboration and sleuthing work. It is also an example of voluntary work that is done for free, as opposed to the work done by Weisenburger and others, which is paid academic work. It is unclear to what extent scholars are likely to cite *Pynchon Wiki* as a source; whereas Weisenburger's companion to *Gravity's Rainbow* has been cited 90 times, *Pynchon Wiki* has been mentioned by 32 sources and cited (with Ware listed as the "author") only twice.[14]

Pynchon Wiki is at the boundary of what we consider digital research: it is essentially a collection of contributions to an online reference book. It fits digital research as we have defined it only in a rudimentary way because only basic computing operations—essentially cataloguing and bringing together digital materials—is performed on the material. Nevertheless, the project fits our definition of e-research insofar as the technology supports the distribution of tasks—a scaling up of the division of labor from a single author to many volunteer author-contributors. And at the margins, computing *does* perform an operation insofar as entries can be linked and searched with data, insofar as that term can be used in the humanities.

Pynchon Wiki is not an example of the scientization of the humanities (which we discuss more later because such scientization is taking place even in literary studies, as in the approach of "distant reading" [see Jockers 2013 and Moretti 2013]). The task of interpretation is not replaced with a more empirical or positivist approach to the text. This wiki *is*, however, an example of the technologization of the humanities in that the contributions are organized by means of the technology: again, unlike in the book version, links are used for cross-referencing, which enables the user to jump from the page on which one entry occurs to other pages and entries. Contributors are able to check how entries have progressed, which was especially important in the early phase when the many contributors needed to gauge when and where they needed to contribute—a kind of "mutual dependence" or "task certainty" (Whitley 2000). They are also now able to check the edit history and contribute updates and so "finalize" the project. Technology has also allowed the addition of images and other digital non-text resources. And finally, on the side of the consumption rather than the production of knowledge, this resource is more readily accessible and provides richer (but not necessarily higher-quality) materials than the equivalent book resource.

This is an example of a research technology that emerged interstitially: it emerged in computer science and can be applied across many disciplines, as we have seen and will continue to see in the case of other wiki-type efforts. *Pynchon Wiki* is a way to enlist "many minds," an approach that has also been applied in the life sciences (Mons, Ashburner, Chichester, et al. 2008). The technology is relatively simple. A single rented server serves as an "infrastructure"—here in quotation marks because the technology is so minimal—for many minds to make contributions.

Note the specific advantages that the technology brings in the Pynchon case: a printed book can be updated or left open to endless revision only with great difficulty (revised editions), but the hyperlinks between page and occurrence of online entries allow research that comes close to computer-supported linguistics (How often does Pynchon refer to x in this novel as opposed to in other novels?). But the main advantage of the wiki form is that many contributors can work in tandem, a kind of parallelization of people rather than of machines. Put differently, *Pynchon Wiki* is a means both to divide and to reaggregate intellectual effort, an advantage that can at the same time be seen as a disadvantage if the outcome is a lack of quality control or consistency compared with what a single author might provide.

Conclusion

In this chapter, we have seen three case studies that are only a sample of the ways in which people and machines are linked together using the tools of e-research. Many more examples exist, but most would be variations on the theme of linking machines together, linking people together, and linking people with machines. Linking can be done in a variety of ways, and the technologies that enable the links to be formed and maintained can be simple or complex, but all draw upon the notion that networks of entities (whether human or nonhuman) can do things that isolates cannot. Computation that would take months or years on a stand-alone computer can be done in minutes or hours by distributed systems. Classification that would require thousands of person-hours for a single laboratory can be done in hours or days by distributed teams of workers. Annotation that would require a devoted, years-long effort by a single author can be done in weeks by distributed enthusiasts. Speed is one aspect, scale is another. What these examples also show is how new roles are emerging and new ways of doing research are congealing into sociotechnical systems (such as life scientists, astronomers, and humanities scholars themselves gaining computer expertise or turning to and working with computer scientists). We also see a great variety of complexity: in this chapter, one case with a high level of computational and collaborative complexity and two cases where computational complexity is low and collaboration relatively straightforward. Finally, we see research technologies providing a focus for intensifying and scaling knowledge production, developing new intellectual directions and drawing users to adopt new practices and engage in the movement to promote computational approaches to research. We have shown various structural transformations in research, which continues to advance as "science in the making" (or "humanities in the making"), even though we have also provided an account of changes that have already taken place in its intellectual, technological, and social organization.

5 Distributed Data

Cultures of Data Sharing

In discussing data sharing, Christine Borgman has pointed out that natural sciences data are typically not very sensitive, whereas data in the social sciences and medicine are often highly sensitive (2005, 21, and 2015).[1] Other aspects of data cut across fields in a different way: whether data have economic value or not and whether there is pressure from funding bodies and scientific journals to make data available (often high in medicine and low in other fields). There has also been a growing awareness of the benefits of increased data sharing and open access to research data among individual researchers (Vickers 2006), editors of scientific journals (Gardner, Toga, Ascoli, et al. 2003), and public authorities (US National Institutes of Health 2003). Yet the problems of data sharing are also well known: researchers are reluctant to give away the material that allows them a competitive advantage, and there are problems with ensuring that data are maintained after having been shared (Carlson and Anderson 2007). Then there are issues around documentation, standards, training, and different rules and laws in different countries (Borgman 2007). Finally, in relation to personal data, there are specific issues relating to the integrity of these data when they are made available to a large and unknown number of users. This issue is in part a technical question of how data can be securely stored in and retrieved from databases, but it has also to do with the ethics of research.

One of the challenges of building collaborative information systems for research data is that many new projects are actually extensions of existing projects, often going back decades, which have an embedded logic and work practices that are highly resistant to change. This resistance to change cannot, however, simply be attributed to conservatism on the part of individual scientists. On the contrary, many of the scientists discussed in this chapter are enthusiastic about the idea of contributing data to larger collaborations in exchange for the additional data that they will in turn have available to them. In practice, however, protocols that are the result of years of cumulative decisions at the local level have resulted in information-storage systems that are highly idiosyncratic and often resistant to federation. Furthermore, broader legacies, such as how researchers relate to the public, play an important role for research involving sensitive data. To illustrate these issues, we look first at two projects that are in very different domains but that nevertheless share similar barriers to building a collaborative infrastructure and then turn our attention to how one country has dealt with sharing data.

One aspect of e-research that we have discussed in previous chapters is the increasing size of collaborations and the increasing complexity of institutional arrangements to support collaborative and computational research. The shift from small science to big science is not new; Derek de Solla Price (1963) charted it five decades ago in work that helped to develop the field of scientometrics. More recently, scholars in computer science have addressed issues of scalability (Simmhan, Plale, and Gannon 2005), and many papers discussing the implementation of Grid-enabled projects have identified scalability as one of the main issues developers have had to deal with (Shimojo, Kalia, Nakano, et al. 2001; Pakhira, Fowler, Sastry, et al. 2005; Zheng, Venters, and Cornford 2011). These discussions of scalability are, however, often focused on large projects such as physics and astronomy Grid-based projects. Yet smaller e-science, e–social science, and digital humanities projects also face issues of scale as they attempt to share data more widely and contribute to larger data sets. One issue raised in this chapter is how legacy data—or data developed prior to the shift to a new technology—can cause significant problems during efforts to standardize and federate data sets. This, again, is not a new issue, and countless scientists are wrestling with this challenge.[2] It is only relatively recently, however, that researchers have begun to pay attention to how small scientific projects negotiate the changes required as they move toward becoming large, collaborative scientific projects (Carlson and Anderson 2006; Walsh and Maloney 2007) and attempt to sustain these collaborations over time (Bos, Zimmerman, Olson, et al. 2007).

The material in this chapter is drawn from three cases: a systematic study of a humpback whale research project involving federating data about the population and movements of humpbacks in the Pacific Ocean; the personal experience of one of the authors (Meyer) as part of a psychiatric genetics collaboration that became involved in contributing data to a large, shared data repository; and research on data-sharing practices in Sweden. Although these cases are in very different scientific domains, they share a number of characteristics, including decentralized decision making, limited data-management expertise, and long-term collections of legacy data that have contributed to the difficulties encountered as the projects have moved from small science to big science. A central common challenge is the tension between the desire for flexibility and innovation in scientific practice weighed against the need for compatible data standards in large-scale scientific data infrastructures.

Marine Mammal Science: SPLASH

Scientists who study whales, dolphins, and other marine mammals use a variety of scientific techniques to gather data pertaining to marine mammal population characteristics and behavior, including acoustics, genetics, and photo identification. In 2006–2007, one the authors (Meyer) studied marine mammal scientists who use photo identification as a main data collection tool (Meyer 2007a, 2007b). The study was designed to understand the ways in which the scientists' work had changed when they switched from film-based to digital photography.[3] The marine mammal scientists' experiences trying to build collaborative scientific infrastructures for studying whales and dolphins' social behavior share much in common with scientists who study humans' social behavior. The characteristics of the animal populations being studied influence the relevant marine mammal scientists' desire to share data collaboratively and to spend time, resources, and effort building infrastructure for the ongoing sharing of data. For instance, some of the dolphin projects in this research focused on relatively small populations of animals (200–500 dolphins) that did not travel widely. Because the animals were located in a small geographic area and were often studied only by a single group of scientists, there was little incentive for the scientists to share the data. In fact, there were disincentives to share because some of the scientists studying these small populations had concerns about others using their data without having gone through the trouble of collecting it, as reflected in the following quote:[4] "Leah: Well, honestly, I'm very protective about it. ... I guess it rather bugs me that I have to do the work, and everyone always asks me for a CD ... it's our scientific study" (anonymous source interviewed by Eric T. Meyer, September 26, 2006, Croatia). Compare this to the approach taken by a large group of scientists studying humpback whales:

Jacob: We knew the success of our project we had done in [location], but also its limitations because it largely funded the contribution and analysis of photographs, but not the dedicated gathering of data. So it very much relied on who was already doing work in certain places. So there were these huge gaps, and we knew [that] to really to answer the questions about population size, trends, human impacts, stock structure, we had to cover some of these new areas. (interviewed by Eric T. Meyer, January 11, 2007, Washington State)

Humpbacks and other whales can travel thousands of miles during annual migrations, and the total population size of humpback whales in the Pacific Ocean is in the range of 15,000 to 20,000 animals. However, prior to the project described here, there was no reliable estimate available for how many humpback whales actually lived in the Pacific. Although research into humpback whales had been going on for many decades, it was done largely by small groups of researchers located at many different institutions and locations, and they could for the most part only record each time an individual whale visited their research area. Scientists hoping to learn more about humpbacks, then, have an incentive to collaborate with other scientists studying humpback whales throughout the Pacific Ocean. By sharing data, they open up the possibility of being able to track individual animals' movements from place to place rather than just recording their repeated visits to a single location over time, and in this way they can build up a more comprehensive view of the entire population of humpback whales. Until quite recently, however, relatively little formal collaboration occurred, and most collaborations were formed based on informal relationships. These informal relationships were often based on common attendance at a university or through shared contacts built through professional conferences and meetings. Recently, however, efforts have been made to build much larger databases, including a project called Structure of Populations, Levels of Abundance, and Status of Humpbacks (SPLASH).[5] SPLASH involved more than 300 scientists working in 50 research groups located in various areas in the Pacific Ocean (Calambokidis, Barlow, Burdin, et al. 2007). Several contributing groups of scientists were included in the research study reported here (Meyer 2007a).

In SPLASH, as in other marine mammal photo-identification projects, researchers use photographs taken in the field to identify individual animals and to track the sightings of the animals geographically and chronologically. The initial efforts to develop the technique of photo identification go back nearly 30 years. Prior to the early 1970s, much of the dolphin and whale research involved techniques that either disrupted the animals' behavior (such as freeze branding) or used dead animals (including necropsies on carcasses of animals that had either died naturally or were killed for research purposes). Increasing public interest influenced by the nascent environmental movement in the 1960s and the Save the Whales campaign of the 1970s helped draw attention to the need to develop less-invasive techniques. More importantly, the passage of the US Marine Mammal Protection Act of 1974 banned most harassment of marine mammals; current research requires special federal permits to be allowed even to approach the animals closely with research vessels to take photographs. Even though there was some initial skepticism about the ability to unambiguously identify individual animals using photographs, the technique is now widely accepted and is practiced by many of the scientists who study these species:

Dr. Lemoine: The original seed of the idea ... came from talking around the campfire. ... It was one of these fun things where ideas come to fruition independently due to synergism and the overall status of the sciences. In the '50s, I don't think anyone would have really come up with that idea. ... I remember telling [a prominent scientist in 1971] about this idea of photo identifying, and he said, "Don't do it. It is not worthwhile. You're barking up the wrong tree. You can't do it, you'll be disappointed. The only way to do it is to catch them and brand them." But, of course, they use photo identification now very successfully. (Meyer 2007a, 138–139)

Using photo-identification techniques, scientists have amassed large amounts of data on thousands of whales, which are collected in individual catalogs of animal images and databases of related information. The problem as it relates to e-research, however, is that the data have been collected by dozens of individual scientists who maintain their own catalogs of humpback whale images, each with its own cataloging schemes, numbering protocols, and databases of associated information. Most catalogs prior to 2003 consist of film photographs in the form of slides, black-and-white negatives, and black-and-white prints. Since 2003, many scientists in the field have switched to digital photography and have designed additional idiosyncratic systems to deal with the digital catalogs, systems that may or may not be consistent even with their own prior catalogs.

Dr. Parrett: It's just too complicated—so right now I have two databases; one on my older data from 2003 back, which was all of the data collected on film, and now I have a new...database that's all the data collected on digital. ... So this spring I'm actually going to [name of location] ... and we have a collaborative agreement where we share data back and forth, and we'd kept it pretty much in the same format except we need to get more on the same page, and we're going to work with their computer guy up there at the end of May and really get our databases uniform. Maybe then ... the data won't be the same, but they'll be the same format. (interviewed by Eric T. Meyer, April 24, 2007, Alaska, by telephone)

Although many in the field believe strongly in the desirability of sharing these data, considerably fewer currently see a need to make local data collection and storage procedures more consistent within the field. One respondent in this study who had spent time considering these issues was also one of the scientists who, for the time being at least, continued to use film-based rather than digital photography:

> Robert: And if you don't have a really good filing system standardized, that doesn't change every time someone thinks it might be better done a different way. So I'm kind of waiting, I guess, to see if it really stabilizes with a naming protocol and a filing protocol that is not going to wander every time someone comes up with a new software for digital pictures. That happens frequently and people send us pictures off a camera and they'll be in files maybe a Canon software or a Nikon one. And you can convert them all to jpegs and fart around with them, but, basically, I don't want to be a film processor. (Meyer 2007a, 235)

Even among the SPLASH collaborators, scientists often continued to work primarily using their established practices, leaving it up to the five SPLASH area coordinators to reformat and rename their contributions to conform to the project's standards.

> Jacob: ... We were not as dictatorial [as we might have been]. Because we were working with established researchers in the area and kind of seeking broad collaboration, many of the researchers maybe started incorrectly with the assumption that people had their own ways to do things that worked and [that we] weren't necessarily trying to force them to do it one way. But partly because of the rapid start of SPLASH, we weren't fully thought out ourselves. ... I haven't fully thought out why some of it was as screwy as it was in terms of experienced researchers, and the one thing I do think about is that they were dealing with the transition to digital as well, and so they had their own system that worked. (interviewed by Eric T. Meyer, January 11, 2007, Washington State)

This comment illustrates a key point we discuss again later: when small scientific projects are faced with sudden and rapid growth by adding numerous noncollocated collaborators, issues of data management and organization often fall by the wayside until problems later surface. This issue reappears in the later discussion of the GAIN psychiatric genetics project. For the SPLASH collaboration, the scientists thought first and foremost about getting out into the field, finding humpback whales, recording the whales' identifying features with a digital camera, and recording data such as geographical positioning system information and environmental data. The timing of the beginning of the SPLASH collaboration also contributed to the confusion. The first year of the SPLASH collection was

2004, and many of the contributors had either switched from film-based to digital photography either in 2003 or 2004 and were still working through how to adapt their methods and organizational practices to the new technology.

The SPLASH collaboration is just one example of scientists who have struggled as their small disconnected projects are faced with a relatively sudden increase in collaboration and scale. In the case of SPLASH, two forces pushed this change. The first was scientific: the scientists' desire to better understand the humpback whales' population structure and long-range behaviors. The second force, however, was economic: SPLASH was a funded project and thus offered scientists the first real chance to begin to respond to these scientific goals. Science costs money, and new funds attract new scientific projects (Carlson and Anderson 2007). This monetary inducement can make the scientific desire to collaborate come into sharper focus for busy scientists with many demands on their time and attention.

This overview of some of the issues facing the SPLASH collaborative shows how a small scientific project can struggle with information issues as it tries to contribute to larger scientific data infrastructures. Next, we turn to a scientific project in a completely different domain to understand how several aspects of SPLASH are not unique.

Psychiatric Genetics: GAIN

The second project that illustrates how small scientific projects can struggle when faced with contributing to larger scientific infrastructures involves genomic research into the basis of certain psychiatric disorders, specifically bipolar disorder (BP). Like humpback whale research, this research is also a long-standing project; over a period of 20 years, researchers have collected blood, genetic data, and phenotypic data on thousands of subjects. Again, although this study at first blush may appear to belong to science, the portion we discuss here also has social science dimensions: much of the data are interview data on the behaviors and social interactions of individuals with certain mental health disorders and of their family members. The data we use for this section were not collected systematically but were the result of a central role that one of us (Meyer) played in the collection and management of phenotypic data for the genomic project over a period of ten years, from 1997 to 2007.[6] During this time, the BP collaboration grew somewhat (expanding from 4 collaborating institutions to 11) but was still primarily an example of small science. Each contributing university had a small number of staff working on the project, usually from one to five staff members, and the entire collaboration involved fewer than 50 people.

In 2006, the BP project was one of six long-term studies in the United States selected to be an initial contributor to GAIN, the Genetic Association Identification Network. GAIN was a public–private partnership project between the US National Institutes of Health and a number of private sector firms, including Pfizer, Affymetrix, Perlegen, and Broad. No funding was offered to studies selected to participate in the GAIN project. Instead, GAIN's approach was to use a carrot to attract scientists to contribute their data: they would get access to extensive genotyping information on their research subjects in the form of genotyping using one million single-nucleotide polymorphism (SNP) microarrays. These one million SNP chips are an order of magnitude larger than the data of many of the previous genotyping projects available to the scientists. Table 5.1 shows the growth both in the collaboration over time and in the size of the data in question. One doesn't need an in-depth knowledge of how DNA sequencing works to understand the scale involved; suffice it to say that the project moved from dealing with hundreds of pieces of data per subject and hundreds of subjects in the earliest period of the study (approximately 10^6 cells of data) to millions of data points for *each* of thousands of subjects (approximately 10^{11} cells of data).

A researcher's contribution of data, however, also had a price: both the phenotypic information that the scientists contributed and the genotypic

Table 5.1

Data Needed to Answer Key Questions in Psychiatric Genetics Case Study in Different Periods

Years	Type of Study	Samples (n)	DNA Sequencing	Scope of Collaboration
1985–1997	Family association/linkage	300	Hundreds of loci/ candidate genes	4 sites in the United States
1997–2007	Family association/linkage	1,500	10,000 SNPs	13 sites in the United States
2007–2009	Genome-wide association	5,000	1,200,000 SNPs	Multiple multi-institution collaborations in the United States
2010–2013	Whole genome	30,000	Millions of SNPs	Worldwide collaboration
Future	Whole genome sequencing	?	Entire genome sequence	Worldwide collaboration

information generated as part of GAIN were to be made immediately available to researchers worldwide, including to pharmaceutical companies hoping to use the information to develop new (and potentially profitable) drugs. One major change for the scientists contributing to GAIN had to do with the embargo period. In the past, data collected by the scientists were typically released to other researchers one year after the final collection of data ended and the data had been cleaned for use. This meant that the scientists had exclusive use of the data in their raw format throughout the data-collection period and for at least a year in analyzable, final format. In the case of GAIN, however, the genotypic information was released to all parties at exactly the same time. The contributing scientists had a nine-month period during which they had exclusive publication rights, but after the nine-month period was up, anyone could publish findings from the data. Although the difference between nine months and a year may seem minor, recall that the previous embargo of a year was for initial access to the data, but more time would then be required to analyze the data. In the case of GAIN, however, there was no embargo at all for access to the data, only for the ability to publish. As a result, the scientists were faced with working on a much tighter schedule and at the same time were analyzing data sets that were several orders of magnitude larger than those to which they were accustomed.

During follow-up interviews with several GAIN participants, all expressed satisfaction with the way the process had worked out but also indicated that they did not feel that they had been "scooped" on their own data. In fact, there was a slight sense of disappointment that the effort had not had more impressive results: two scientists who had initially been extremely optimistic that the project would result in major scientific breakthroughs said that in the end the results were more modest than they had hoped. As of 2011, 87 published papers[7] cited the collections in the GAIN study, and 32 papers[8] were published by investigators in the GAIN project. Although this number of publications is respectable, it is not indicative of the breakthrough that some had expected.

One hurdle that the BP project had to overcome after being selected as one of the initial six GAIN studies was that the subjects being included in the genotyping had been interviewed over a period of 20 years using three different versions of the interview instrument, which in turn were encoded into three different phenotypic databases with incompatible variable names and formats. The largest set of items in these phenotypic databases consisted of answers to more than 100 pages of questions, administered as semistructured interviews performed by trained clinical researchers.

Interviews took from four to six hours on average to administer and resulted in recorded values for approximately 2,600 variables. In addition to these data, tables were used to record each research participant's "final best estimate"—the clinical diagnosis assigned to that person based on a trained clinician's analysis of his or her interview(s), family history, medical records, and other information. Multiple best estimates were made for each subject because at least two clinicians plus the interviewer and an editor assigned their own diagnosis, but each subject was given only one "final best estimate." These final best estimates are in the form of a hierarchical diagnosis using diagnostic systems that changed over the years of the study. The earliest diagnoses used a combined DSM-IIIR/RDC system, and the latest subjects were diagnosed with DSM-IV.[9]

Because this interview schedule went through several iterations over the 20 years of the project, there are three main versions of the phenotypic database. The first includes data collected via paper interviews and entered into an Oracle database designed and maintained by a US federal contractor. The second set of interviews were also completed on paper and then entered into a Paradox database designed by a database developer located at one of the project sites. The third set of interviews were initially done on paper but were then transitioned to direct-entry interviewing via laptops and tablet PCs using a proprietary database designed for the study by an external company. All three versions have been converted from their native storage formats into Statistical Analysis Software (SAS) files for use and analysis, but their variable names are not consistent. For instance, the same variable might be designated "I1120" in the first set of variables, "Number_of_manic_episodes" in the second set, and "V756" in the third set. Although this variation seems confusing to outsiders and makes combining data difficult, those familiar with the data have found that the use of very different names serves as a quick way to see at a glance the source of a variable or set of variables.

The differences in variable names illustrate the difficulty in combining data from several iterations of the same project, let alone trying to combine that data with data from other projects. Because the decisions regarding things such as variable naming conventions was left to database designers rather than done in a systematic fashion (and no standards exist for these data), trying to combine these data later requires a fairly high degree of understanding of the research project. One of the contributing sites had several staff members working on a combined data set that would convert variables from all three versions to a standard naming system; this project took more than two and a half years to complete. Also, because the interview schedule changed between iterations and questions were rewritten, added, and deleted from version to version, there is no clear mapping from one to another in the majority of cases. When the data were used primarily internally by people very familiar with the research, the analysts were able to informally share knowledge about how best to use the data. When such data need to be shared more widely, however, these idiosyncrasies can be very confusing. In addition, the group in charge of GAIN data distribution also required well-documented data dictionaries for the databases, which had been kept throughout the project, but not in a format compatible with the GAIN requirements for submission.

This case illustrates how decisions made by a wide variety of people over a period of many years can have major implications when the scientific data are later reused in ways that the original designers were unable to foresee. The accumulation of many small decisions, most of which were sensible at the time, can subsequently result in considerable work trying to reconcile the many differences that are the result of those decisions. Put differently, the creation of a coherent digital resource from across multiple sites and with an increasing scale and scope, including from databases that precede e-research, requires integrating multiple data sources and organizational practices that also leaves in place some of the idiosyncrasies built up over time.

Data Sharing in Sweden

Data collection is particularly sensitive in Sweden since public authorities regularly gather large amounts of data about their citizens.[10] As we shall see, concerns about the privacy and anonymity of data come into focus especially when there are breaches of law or accepted norms. Sweden has a number of well-established data collections and registers in addition to a system of national identifier numbers linking individuals to information. These collections and registers provide an essential precondition for e-research, but sharing such powerful records digitally also presents greater possibilities for abuse. Anne-Sofie Axelsson and Ralph Schroeder (2009) carried out a study to assess how key actors, mainly database owners or managers and database users, have addressed the more powerful risks and benefits of digital research data. The study also examined the debates in Sweden about research data and public attitudes. As we shall see, Sweden sheds interesting light on the possibilities and limits of data sharing.

Data about individuals constitute a special case because these data are governed by different laws and regulations regarding the collection, storage, use, and sharing of data about individuals in different countries. Sweden presents an interesting case study for data sharing because it has rather unique social circumstances for gathering and making data about people available to researchers. A sketch of the Swedish e-science program's background puts this uniqueness in context.

Sweden, like other countries (as discussed in chapter 3), has developed an ambitious e-science program. As in the United Kingdom, the term *e-science* has been adopted in Sweden.[11] The Swedish e-science initiative can be seen as consisting of a national Grid infrastructure, a university computer network, and a well-established but underused set of databases and registers. Its development as a program began in 2005 when the Swedish Research Council established the Committee for Research Infrastructures with the remit to support the building and use of infrastructure for Swedish research (see Vetenskapsrådet 2009). A year later, in 2006, a new committee was established by the Swedish Research Council, the Database Infrastructure Committee (DISC).

At that time, there were two already existing parts to the research infrastructure: the Swedish National Infrastructure for Computing and the Swedish University Computer Network. Together with these two already existing parts, DISC was, according to the Swedish Research Council, supposed to "comprise the trinity which constitutes the base in the work of the Swedish Research Council with an e-science infrastructure for Swedish Research" (Vetenskapsrådet 2009). The establishment of DISC put the focus firmly on databases and data sharing within the Swedish e-research enterprise because DISC's general mission was to "create an advanced coordination of existing and new quality-assured research databases and provide this national resource to Swedish and international research" (from the DISC website http://www.disc.vr.se/, accessed April 2009). Data in this case mean mainly data in the social and medical sciences, whereas in other countries various e-research programs cover both broader and narrower types of data and e-research.

In December 2008, the Swedish Research Council announced a program of funding for six database research projects and one postgraduate school, totaling 135 million Swedish kronor (approximately US$20 million) for a five-year period within the Swedish Initiative for Micro-data in the Social and Medical Sciences. This initiative's aim was to fund a small number of database research projects per year, and these projects, led by junior researchers, would be joined together in a network with joint workshops, conferences, and a postgraduate school. The vision for this part of the enterprise, the Swedish Research Council said, was to create a new generation of researchers with expertise in using Swedish databases and registers and knowledge about database methodology and with start-up funding for new and interdisciplinary database research (Vetenskapsrådet 2009). In the autumn of 2008, the Swedish Research Council, via DISC, established a national service, the Swedish National Data Service, which has since then been responsible for maintaining existing databases, primarily within the social and medical sciences, and supporting the creation of new ones. Its remit has since been extended to all types of data, including data in the humanities. In short, Sweden has undergone a major overhaul or upgrading of its data infrastructure for research, pushing it in the direction of e-research—an upgrade that continues today.

Sweden has a long-standing and unique collection of databases and registers (Welin 1990; Jonsson and Landegren 2001), including a number of official registers dating back to the eighteenth century (Demographic Data Base 2009) and current ones collected by the official body Statistics Sweden (2012). The uniqueness of the Swedish data collections and registers and the national conditions that relate to them stem from the following factors unique to Sweden (see also Jonsson and Landegren 2001):

- Many of the registers have been built up over a long period of time. The first census registers were established in the eighteenth century on the basis of parish statistics. These registers can support genetic studies by providing information about the relatedness of the sampled individuals.
- Sweden has continually had a large number of official data registers for some time, which are maintained by Statistics Sweden, the Swedish central-government authority for official statistics. Statistics Sweden has carried out population-based surveys (including the whole Swedish population) for some time, thus guaranteeing a high degree of reliability of the survey data.
- The Swedish health-care system has functioned well for a long time and is highly advanced, so there are extensive data about individuals in medical registers.
- Sweden has a low population, which increases the likelihood that two individuals who have a disease also share the genetic risk factors—an advantage in genetic studies.
- Since 1947, Sweden has had a system whereby every newborn or immigrant to Sweden has a national identification number, which is used by all public administrations and in most commercial contexts—including economic transactions (see Otjacques, Hitzelberger, and Feltz 2006 for a European comparison). This identifier can be used in research for cross-referencing between registers.

- Finally, a number of data collections and registers have been built up within individual research projects, and they are available to other researchers via contact with the managers of the individual data collections. To mention a few prominent examples, there is the Swedish Twin Registry (described in detail in Lichtenstein, De Faire, Floderus, et al. 2002), the Stockholm Birth Cohort Study (described in Stenberg and Vågerö 2006), and the Demographic Data Base (described on its website, Demographic Data Base 2009).

These unique resources provide part of the rationale for the Swedish Research Council's drive to provide greater access to these national data resources (Vetenskapsrådet 2005, 5). The council points out that these data collections are highly underused because, first, many researchers are unaware of the existence of the collections and registers, which are stored at and available only via particular institutions, and, second, because it is considered complicated and costly for researchers to access them (Vetenskapsrådet 2005, 27).

A description of data collections by the Swedish Research Council (Vetenskapsrådet 2005, 9) distinguishes between registers at official authorities such as Statistics Sweden, registers at universities and colleges, and registers at hospitals and health-care institutions. Swedish official statistics are collected, stored, and made available by 25 different official authorities, each responsible for its specific area (e.g., higher education, health and welfare, agriculture). All of these official authorities belong to the network Official Statistics of Sweden[12] led by Statistics Sweden. In addition, Swedish universities and colleges have also built up a number of extensive and unique social science and medical longitudinal databases (Vetenskapsrådet 2005, 13).

The public's trust is an important precondition for collecting and keeping personal data. At the same time, the public also has much to gain from research based on information about themselves. In Sweden, the public has for some time trusted the authorities in relation to personal data. Surveys of public attitudes toward collection of individual data show that people are positive toward giving out information about themselves to the Swedish authorities. For example, in a report from the Ministry of Justice (Justitiedepartementet 2007) about issues related to integrity in Sweden, 39 percent of respondents ($n = 1,000$) stated that they think that the authorities collect a reasonable amount of information about its population, and 12 percent said that they collect too little information. Only 10 percent of the respondents thought that the authorities collect too much information.

Regarding the national identification number, a majority of the respondents, 66 percent, thought that it is used to a reasonable extent, and only 1 percent thought it should be done away with.

Nevertheless, over the years a number of incidents have caused headlines in the Swedish newspapers, heated discussions in the media, and even demonstrations. One well-known incident (described in detail in Welin 1990) took place in the 1980s, whereby the largest newspaper *Dagens Nyheter* (1986) revealed details from a large social science study of 15,000 individuals in Stockholm, all born in 1953. The study had been ongoing since 1966, and the researchers, the newspaper revealed, had collected qualitative as well as quantitative data about the individuals, in part from population-based registers and without the individuals' knowledge. The Metropolit study, as it is known, caused a huge debate and even demonstrations against the way researchers could make use of information about individuals without the individuals' knowledge, for undisclosed purposes, and with financial support from the state. Without going into detail about the study and the reactions, we can note here that this incident—as Göran Hermerén (1986) pointed out at the time—initiated the first open discussion of research ethics in Sweden and produced a strong negative reaction against the researchers' "secret" activities.

Another more recent case in Sweden that caused not only an uproar in the media but also became a legal case is the so-called Gillberg case. A professor in child psychiatry, Christopher Gillberg, was ordered to hand research material to another researcher for inspection but instead destroyed the material to protect the integrity of the study participants (children diagnosed with attention-deficit hyperactivy disorder and their parents). For his action, Professor Gillberg was conditionally sentenced and fined by the court of appeal in February 2006, though the case subsequently also made its way through the European courts.

Another more recent debate in Sweden has had to do with the new National Defense Radio Establishment law, a set of laws that allows the National Defense Radio Establishment on commission by the Swedish authorities to gather radio signal intelligence from data traffic (Internet and telephone) over the borders to and from Sweden. This law was proposed in the spring of 2007 and approved by the Swedish Parliament in the summer of 2008. It was heavily criticized and widely discussed by politicians, the public, and academics in political forums, in blogs, and in Swedish media before and after the decision. Strong criticism finally led to a number of changes in the law, the main change being a stronger protection of the individual's integrity. The law finally came into force on January 1, 2009.

What can be said about these three incidents—each causing, at different times and in different contexts, a huge debate and uproar among the public as well as within the scientific community and among politicians—is that they touched upon several of the issues that are at the very core of e-enabled data sharing. In the Metropolit and the Gillberg cases, the public, other researchers, and the media accused the researchers of not being open about their activities. However, there is also a difference between the two cases. In the Metropolit case, the lack of openness was due the fact that the researchers were not aware of (or ignored) the concerns that the public and especially the study participants might have had about having information about themselves stored in computers and shared among researchers. Moreover, Swedish legislation did not prohibit this way of using the data or conducting research. Nevertheless, due to the public reaction, further studies on the material were prohibited without the consent from the study participants. In the Gillberg case, in contrast, the lack of openness was a way for the researcher to protect the study participants' integrity. In this case, however, the researcher was judged (in the Swedish ruling) to have acted incorrectly according to the law. Nevertheless, further research on the data and perhaps new knowledge regarding the disease that was being studied were made impossible. The most recent case, the National Defense Radio Establishment law, also has a clear bearing on e-enabled data sharing in Sweden because the law and even more so the debates that preceded it made Swedes worry that the (Swedish) state may want to collect and use information against its own citizens.

Several lessons can be drawn from the Swedish situation outlined here. The people we interviewed about it[13] are well aware of the unique possibilities in Sweden:

> We have had unique conditions here in Sweden to conduct this kind of [longitudinal population-based] research. Among other things, we have a well-functioning national registration from which we can select a certain subgroup, and we can follow these individuals wherever they go and move. Our American colleagues would go mad if they even tried to get the same kind of data input to work. (interview with database manager and database researcher)

This researcher also phrased this point in terms of "competitive advantage": "In Sweden we have a niche in this, to conduct longitudinal population based research, which is a competitive advantage in comparison to other countries, based on the fact that we can collect data of this kind."

Interviewees also recognized the dangers of the Swedish situation, particularly in relation to sensitive information: "In Sweden we have more problems with secrecy compared to other countries, because our surveys are complete while in other countries the quality of the sources are too low, they have only selections" (interview with database manager). At the same time, in the event of collaboration with international colleagues, the informal relations of trust would need to be extended in studies requiring shared data. As one researcher using longitudinal data put it, "The study was conducted in collaboration with American colleagues, and no single individual had been capable of carrying out this on his own. ... The collaboration was not the least initiated or administered from above but totally based on trust, on personal contacts, on the fact that people liked each other" (interview).

Trust would need to extend into the e-research environment: "The Swedish legislation is very rational, both for researchers and individuals. It takes into account the needs of the society as well as the needs of the individual for protection. ... The question of trust is the key!" (interview with database funder and database researcher). Similarly, the funding council and lawmakers would therefore need to tackle problems arising with data reuse: "Swedish legislation does not allow reuse of data...which is something the Committee for Research Infrastructures *really* should try to do something about. ... The whole idea with Swedish National Data Service builds on this, but the legislation does not allow it" (interview with database manager).

Other issues recognized outside of Sweden were also mentioned in these interviews, such as the short funding cycles, which may mean that databases do not come into widespread use or the focus will be on coming up with new data rather than on reusing existing data: "People move in and out of studies. What is it that ensures the continuity of a database? A lot of existing databases are of high quality but have fallen into oblivion. That is a waste of resources. But the research funding system tends to reward collecting data above using already collected data" (interview with database manager and database researcher).

The stakeholders we interviewed seemed to worry little about storage capacity or which middleware to use to make data available and much more about how the exceptional Swedish conditions that have been built up over a long period can be preserved and protected from misuse that would ruin them for research. In other words, they were aware that as Sweden tries to build up an infrastructure for e-research around its unique data collections, the key will be also to maintain the unusual conditions of trust that have obtained in Sweden. As we have seen, a number of incidents related to Swedish data registers and data sharing have affected the public's trust in the state and in researchers and shown that this trust is fragile and can easily be destroyed.

Significantly, the two main overarching issues that the interviewees in this study stressed were openness and safety: openness with the public and within the scientific community regarding what researchers are up to, and safety for—or protection of—the individuals that contribute information about themselves to the data registers. The uniquely high level of trust that has so far governed the relation between researchers, the public, and a social order in which personal data are thought to be safely and transparently managed will face challenges as it is extended into a new environment. Nevertheless, the social preconditions that govern data sharing in the Swedish case are bound to continue to shape it. Other countries obviously do not share the same setting of trust and would need to build on a different relation between the public, researchers, and the way in which personal data are governed.

Conclusion

As we saw in both the SPLASH and GAIN collaborations and in the Swedish case, the shift from small to big science can be challenging. Most of the personnel working on SPLASH and GAIN were trained in scientific methods and theory, but few participants had any systematic background in data management and organization, and even with the national efforts in Sweden, new institutions and data-management roles within these institutions have had to be created. In the case of SPLASH, all of the personnel responsible for designing the database systems and methods of information organization were trained in biology, and none had any formal training in database design or information management. The decision regarding which personnel to assign to these duties relied primarily on identifying staff members with an affinity for and skill with computers. Although the databases that were designed as a result were perfectly useable, they did not incorporate fully normalized designs or other features that more trained designers might have included. More importantly, because they were designed by a single researcher or small group of researchers, they will most likely be incompatible with other databases—for instance, if SPLASH wishes in the future to federate its data even further (possibly by expanding to other regions or by incorporating additional species of whales).

In the case of the GAIN project, a small number of people with considerable expertise in data management were part of the decision-making process, but even in this case many of the decisions were not made systematically. For instance, the format for variable names for the third iteration of the phenotypic interview was decided by the company that programmed the database and had more to do with the structure of the company's particular implementation of an entity–attribute–value database design than with the scientific analysts' needs. Even the second iteration of the interview, which was stored in a database designed by a skilled analyst, experienced unexpected confusion in the naming of variables when a number of names including the ampersand (&) were altered when the data were imported into SAS; SAS does not support the ampersand and so converted both that character and any spaces to an underscore. Thus, "Total Manic & Depressive Episodes" in Paradox became "total_manic___depressive_episodes" in SAS, with three consecutive underscores in the center.

Because much of the analysis in the past had relied on ad hoc requests for subsets of the data made by investigators to the small number of data experts working in the collaboration, much of the knowledge about the idiosyncrasies of the data sets was never written down in a systematic fashion. When GAIN required that the data be shared and documented, considerable effort had to be made to translate this knowledge into a written format. Also, although the data sent to GAIN were cleaned, the three versions of the data set were still separate. One research group's internal effort to construct a unified data set was not yet finished at the time the data needed to be provided to GAIN, and the group that had spent so much time and effort on combining the data was unwilling to release the combined data set publicly until they themselves had gotten use out of it. The plan was to release the combined data set to the scientific community after a year of internal use.

Among the striking similarities between these very different scientific domains is the extent to which existing practices are the result of lots of small decisions made in a number of small research projects by many individual researchers and technicians over a long period of time. These decisions were often made with little or no discussion of the impact beyond the particulars of the specific local study, often because scientists at the time did not anticipate the future need to share the data with other scientists. As a result, both SPLASH and GAIN had over a period of years adopted highly idiosyncratic methods that hindered an easy transition to sharing data. Scientists therefore now find that they have to spend a great deal of time, effort, and money to transform their data into forms that can be used in a larger data-sharing project. In some cases, these barriers may be high enough to dissuade scientists from contributing at all. In other cases, unless the idiosyncratic nature of the data is diminished, users of the collaborative data may find them confusing or misleading. These social realities of scientific practice must be addressed if e-research projects are to be successful, particularly when applied to existing scientific protocols.

In Sweden, in contrast, many of the people we interviewed were professional data experts; the process of moving to e-research could be built on established institutional infrastructures; and the expertise contained in these infrastructures made for an awareness both of the possibilities of databases and of their limitations. These limitations include some technological limitations, but others are primarily social. For instance, databases risk becoming underutilized even when they are of a high quality.

The data developed such high levels of idiosyncrasy in the whale project and the BP project because research in the fields in question had historically engaged in smaller-scale collaboration and because the new project collaborations were also being done in a very noncentralized fashion (which again contrasts with the Swedish case). The quote from "Jacob" cited earlier in the chapter mentioned that SPLASH did not "try to force them to do it one way." Likewise, the BP project that was part of GAIN was always very decentralized, to the extent that individual contributing sites were able to choose to skip portions of the interview schedule and choose to use alternative systems of organization and management locally as long as they contributed their final data in the agreed upon formats. Both the SPLASH and GAIN projects had nondogmatic leaders who were flexible in their approach to managing the collaborations. They by and large did not attempt to impose decisions but instead sought consensus and allowed considerable individual latitude to their colleagues and contributors. This flexible and decentralized form of leadership is common among scientific and creative teams (Mumford, Scott, Gaddis, et al. 2002) and is not inherently problematic. Science relies on scientists' freedom to innovate (Bush 1945; Gordon, Marquis, and Anderson 1962), although some recent work suggests that these patterns are changing in the face of calls for measures of increased accountability in and relevance of scientific work (Demeritt 2000; Harman 2003). Having collaborative science designed with wide latitude for individual contributors to pursue unique contributions can arguably lead to better, more innovative science. From a data-management perspective, however, lack of reliance on standard data structures, naming styles, and metadata (data about data) makes federating data either difficult or impossible. These challenges are something that a national effort, such as the one in Sweden, can avoid in the way that some individual projects cannot. National infrastructures are run by bureaucracies, and bureaucracies require standards, rules, documentation, and clearly defined lines of authority; the typical features of Weberian bureaucracies also often govern large-scale research collaboration (Shrum, Genuth, and Chompalov 2007).

One additional issue at the project level is that data management often is not considered a top priority during the startup phase of individual scientific research projects. In the cases described here, few of the scientific decision makers had detailed knowledge of the demands of data management and as a result treated it as ancillary to the main scientific research design. Also, in both the SPLASH and GAIN cases the studies were rather hastily implemented and saw a number of operational changes during the early phases. Encoding these changes into data systems became a case of trying to hit a moving target until the scientific protocol had stabilized. Based on our observations of these and a number of other scientific projects, however, this initial uncertainty is not uncommon with grant funded research. In the United States, grants are written, submitted, and revised over a period of years in many cases, and by the time funds are secured, local changes in personnel and wider changes in the state of current scientific knowledge have occurred. The grants are also written with some flexibility in terms of specific activities, and the decisions about how to implement the research are often left until the funds have been secured.

Borgman (2015) argues that e-research is putting data management and reuse at the top of the agenda in research policy, although this shift is recent and applies here mainly to the still ongoing policies at the Swedish National Data Service. The question will remain as to what extent data-management demands should dictate scientific decisions or if individual scientists should decide on issues of compatibility and data availability. This issue will continue to be debated among researchers and policy makers working toward large, federated data sets, and it is unlikely that a single answer will resolve this tension. Collaborative scientific projects will continue to balance the needs of individual scientists for flexibility in their data-collection protocols with the demands of federated databases for data to be organized in a consistent and structured manner.

In table 5.2, we highlight several features of the data in all six of the cases we have discussed so far—SwissBioGrid, Galaxy Zoo, *Pynchon Wiki*, GAIN, SPLASH, and the Swedish National Data Service. It lists the types of data each case is dealing with, the origins of those data, the uses of the data, how the data are typically organized, the manipulations performed, the tools used, and how e-research and computational approaches add value to the data.

Table 5.2
Features of Data in Six Case Studies

	Type of Data	Method of Contribution	Uses	How Data Are Organized	Manipulations	Tools	e-Research value added
SwissBioGrid	Potential drug targets	Data from drug company	Analysis of fit of targets	Databases	Matching	Middleware	PC power; speed
Galaxy Zoo	Pictures of galaxies in, classifications out	Data from SDSS; eyeballs, Internet access, volunteering	Typical: taxonomy; atypical: Green Peas and Hanny's Voorwerp	Web backend database	Classification	Website	Manual labor in support of digital analysis; speed; creative new discoveries; organizing power
Pynchon Wiki	Descriptive words that explicate something in the novel	Any registered user can contribute, discuss, and revise	For reference	Cross-referenced hyperlinks (page and item)	Search, linking	Wiki	Multiple sources; increased speed; organizing power

Distributed Data

Table 5.2
(continued)

	Type of Data	Method of Contribution	Uses	How Data Are Organized	Manipulations	Tools	e-Research value added
GAIN	Contributors' phenotypes	Answering questions structured to support DSM algorithmic approaches to diagnosis	For analysis, in combination with genotypic data	Relational databases	Turning lives into digital data, then statistics	Statistics software	Additional statistical power
	Genotypes from blood samples/cell lines	Drawing, freezing, and transforming blood samples, then genotyping those samples	For analysis, in combination with phenotypic data	Flat file databases	Turning blood into digital data, then statistics	Genotyping machines; statistics software	Additional statistical power
SPLASH	Photos of whales	Consolidating photos centrally	Identification of numbers and movements of whales	Relational databases, cardboard boxes	Matching	Photo software and simple databases	Sample size
Swedish National Data Service	Health and population records, among others	Merging many databases via the primary key of "person number"	Health and social research	Relational databases	Turning health and government records into research data	Virtual research environments, research software	Ability to link multiple data sources due to ubiquity and coverage

Data, information, and knowledge are typically viewed hierarchically, with each subsequent level representing a higher-order operation than the previous (Henry 1974; Zins 2007). Raw data require cleaning, organization, manipulation, and analysis before they can transition from data to information and knowledge. In the academic world, knowledge is communicated through publication, but data sets also are often published and include metadata that make them usable to others.

At a certain level, data consist of atomized (nonreducible) entities that belong to the objects or phenomena under investigation, are open to checking (they cannot be immunized against verification or falsification), and have a particular relation to the natural or social worlds "out there" (recall that we use a "realist" definition of science). Thus, when data are used to represent and intervene, we say that they are scientific, and otherwise we say they are research materials (or "data" in a discipline-specific sense). The manipulation and organization of data in sociotechnical systems—that is, e-research technologies—requires considerable effort, as we have seen.

Nevertheless, data and their manipulation contribute to the advance of knowledge because they provide powerful inputs about the world into research, mainly because the scale and scope of digital data are typically larger. Data come "before" their organization: as Ian Hacking puts it, the view that "all data are of their nature interpreted" is misleading: "data are made, but as a good first approximation, the making and taking come before interpreting" (1992, 48) (He adds, "It is true that we reject or discard putative data because they do not fit an interpretation, but that does not prove that all data are interpreted" [1992, 48] and goes on to distinguish data from other parts of the scientific process, such as the calibration of instruments.) Nevertheless, data, once recorded, require schemata to organize them. The sensors that record temperature, for instance, are themselves subject to variations based on environmental conditions, and knowing the margin of error that this variation entails and the conditions in which the sensor operates are useful pieces of metadata that enable more accurate transformations of data to information to knowledge.

By way of example, one of the technology advances in recent decades is the availability of small and cheap sensing devices that can be deployed in the field in various settings to provide streams of data back to scientists operating in the lab.[14] CENS was set up to explore how these systems could be deployed and to discover new uses for remote sensors (Borgman, Wallis, Mayernik, et al. 2007). As noted in chapter 1, one of Borgman and her colleagues' findings was that the data generated by sensors were viewed differently by participants from different domains, with engineers interested in performance data from the sensors, scientists interested in scientific data, and data managers interested in contextual data that could influence the scientific data. Thus, the information that made data analyzable may have been added at the point of sensing but was of varying interest to different partners of a scientific team.

In the six cases examined here, we can notice that different processes govern data organization and manipulation and how each process adds value to research. For SwissBioGrid, scaling up computational power resulted in increased speed of analysis and thus made the discovery of potential drugs and clinical treatments more effective. Although the data, databases, and methods of analysis were not radically altered, parallelization of machines made the analysis more efficient and large scale.

Galaxy Zoo also increased the speed of analysis, but in this case via the very different mechanism of parallelizing the efforts of human volunteers. It is interesting to note that one of the reasons that Galaxy Zoo works is that the task being performed by the volunteers has been successfully transformed by the platform's designers, from one that, in Richard Whitley's (2000) schema, would yield high uncertainty if performed by people who are mostly not professional astronomers ("What kind of galaxies are in this image of the sky?") to one with much lower uncertainty ("In this single image of a galaxy, does it have spiral arms?" "If yes, then are they turning clockwise or counterclockwise?"). The Galaxy Zoo volunteers are thus in a sense asked to be part of an algorithm that follows a simple logic but needs human input at various points in the algorithm to tell it whether to branch in the direction of one type of galaxy features or another.

The typical use of these Galaxy Zoo data are to create a taxonomy of galaxies that can be used for other, higher-order analysis among the professional astronomy community. However, the system's affordances also allow for atypical uses, such as when citizen scientists, the nonprofessional volunteers, are able not just to classify objects into the known taxonomy but to participate in the discovery of new kinds of galaxies. Both Hanny's Voorwerp (Lintott, Schawinski, Keel, et al. 2009) and Green Pea–type galaxies (Cardamone, Schawinski, Sarzi, et al. 2009) are examples of how the distributed model for classifying galaxies has enabled new and completely unanticipated discoveries to occur. Both are new types of galaxies, never before seen or classified, that have since their discovery been given considerable scientific attention.

In the case of *Pynchon Wiki*, speed and organizing data are again two themes. As with Galaxy Zoo, which uses a relatively simple website as the main interface to citizen scientists, the technology being put to use here is simple: basic wiki software with no particularly novel added computational power. The data in *Pynchon Wiki*, however, are very different from the data in the previous examples: here the raw data, as it were, come from a work of fiction and the facts or interpretations (notably not data in the scientific sense) related to what items that work refers to. The work of the platform is to enlist enthusiast volunteers' abilities, as with Galaxy Zoo, but the algorithm is simply a list of the pages in the novel, and volunteers are put to work identifying what is being alluded to in phrases and passages of interest and then producing text and linking these phrases and passages to information about their meaning, referents, or sources. There may be higher task uncertainty here in terms of what a given passage means, but there is low task uncertainty and high mutual dependence for the *overall* task insofar as the task is for a group of people to work their way through all the pages of the novel and create an annotated version. The contributors were asked to perform higher-order interpretations of the data, but within a well-understood framework that has a preexisting analog counterpart for the participants to use as a model for their work.

In the GAIN case study, two distinct sources of data needed to be linked and combined in order to be useful. On the one hand was the genotyping information, which was generated by extracting DNA from human blood and measuring which genes or DNA segments or nucleotides were in each sample, depending on the exact technique used for genotyping. These data are what most laypersons think about when they think about genetic research: converting the facts of DNA structure via sensors into data and information for analysis. These genetic data are potentially large (when millions of SNPs or whole genome-sequencing techniques are used), but they are also relatively uniform to deal with. In other words, there are not many *types* of data in a database of genetic information, but there are many data points. The transformation here is to take blood and via a series of steps in the laboratory turn it into analyzable data. These data have de facto standards insofar as once scientific teams decide which genotyping chips and machines to use, the manufacturer's formats and protocols generate standard output that can then be shared. A complication here is that the data from later studies are often not fully backward compatible because the locations being measured with a one million SNP chip, for instance, are not a simple superset of those processed with a lower number of SNPs. This incompatibility results in the need to regenotype samples when analyzing them as part of a bigger collaborative project, which was in fact one of the reasons for the GAIN effort: most of the samples selected had been previously analyzed at lower resolutions but were chosen for new analysis based on their potential interest, largely based on the second type of data, which was phenotypic data.

Phenotypic data, as discussed earlier in this chapter, are more complex and messy, even though they typically require a smaller database. Phenotypic data are collected through interviews with human respondents (who may or may not in all cases be reliable witnesses to their own lives, in particular those subjects who have periods of serious mental disorder, which is often complicated by substance abuse). Participants' self-reports are combined with medical records and other evidence to come up with a final best estimate of their diagnosis (for mood disorders in the particular case discussed), which reduces all the phenotypic data down to a single data point with half a dozen or so possible values (such as bipolar I, bipolar II, unipolar, etc.). However, reanalysis is increasingly being done using the full phenotypic data set, particularly when scientists are interested in subphenotypes or in specific patterns of data extractable from the interviews and self-reports (such as the aggravating factors of various types and degrees of substance abuse).

The phenotypic data can also be discovered to be problematic when combined with genotypic data. A fairly common example is discovering that two siblings in a sib–pair study or a parent and child in a family study are not in reality related to one another or are only half-related (i.e., two participants who think they are full siblings, but the genetic data shows that they have different fathers). Because this information (gleaned from the combination of two sources of data) is obviously highly sensitive and potentially harmful to the participants and their families, firewalls are commonly set up between the teams who gather the phenotypic data and the teams who work with the genotypic data. For instance, names and identifying information required to be kept for purposes of follow up are never available to the genetic analysis teams, and the information about genetic relationships is removed from the study and never communicated to the phenotypic research team to avoid any risk that the information will get back to participants. In the BP case described, there was a single lookup list, held securely and with very limited access, that allowed deidentified phenotypic data to be linked to genetic results. However, questions have been raised (not in the context of this study, but more generally) about whether there is any meaningful way to promise anonymity of genetic data once the cost of completely genotyping individuals is low enough to become a commodity. Because a person's genotype is even more personally identifying than a name or a fingerprint, can any promise of anonymity realistically be made if genetic information is stored in accessible databases? These issues also apply to Swedish social and medical records. In any event, the GAIN project, by means of sharing data, was able to enhance the scale, scope, and statistical power of the data analysis.

In the case of SPLASH and the humpback whales, the data under consideration are photographs of whale tail fins, or flukes. The data relate to facts in the natural world—whale flukes and, at an even more fundamental level, the photons of light bouncing off of them and subsequently being registered on photographic devices. The photographs, then, are the raw data, but these data are imperfect: the lighting conditions may have been poor, or the sea may have been choppy, or the boat on which the person taking the photograph stood may have been off to the side of the fluke rather than at the preferred perpendicular angle. Likewise, the ability to match a given fluke to the flukes in the identified catalog vary: some flukes have more distinctive patterns than others, and some animals are even "celebrities" among the researchers because they are particularly interested in humans and likely to approach boats frequently (and thus be recorded more frequently).

The systems for organizing these photographs range from digital databases of photos with features for sorting and categorizing individuals to simple cardboard boxes with small printed photographs of whales held together in packets with rubber bands. The digital photos are obviously more suited to sharing and distributing and open up the possibility of understanding population-level questions about whales. They represent a major change from how research was done in the past and a potential step change in the kinds of questions marine biologists will be able to answer about oceans and ocean life.

The whale data present an unusual added twist for scientists to cope with: whales live a long time. Their exact lifespan is unknown, in part because the ones studied have lived longer than the scientists who have been actively studying them. Estimates range anywhere from 50 years to 200 years, with at least one individual baleen whale accurately recorded to have lived at least 130 years based on the identification of a harpoon point lodged in its body as a type that was no longer used after the 1880s (Gardner 2007). What this means for the field of marine mammalogy is that the data being collected by scientists today are potentially useful for many generations of scientists to follow—not just in the general sense that science builds on previous discoveries, but in the specific sense that marine biologists of future generations will be studying the exact same whales that their predecessors studied. This factor raises a question for the field of how to hand down these data from generation to generation as the objects of scientific study outlive the scientists who study them. Even though nondigital photographs will provide an important "legacy" system, the transition to digital photography enables greater sharing of images.

Finally, the case of the Swedish National Data Service deals with data that, like the whale data, are useful for many generations, but at the aggregate population level more than at the individual level. Because of the availability of a unique number assigned to each person in Sweden, individuals can be tracked from database to database, and their data can be combined for health and social research. The ability to link and combine these data is unique to Sweden even if other countries gather similar types of data: in many places, such as the United States and the United Kingdom, efforts to have single identifiers that track individuals at this level of detail have been resisted. One of the key elements in Sweden is the public's trust in the bodies that hold the data (conversely, in other countries the public distrusts the companies and government agencies that would hold the data). However, Sweden's case raises broader questions about "big data," especially data held in the private sector and used for research (Savage and Burrows 2007, 2009). So having your Swedish doctor know that you are pregnant and your past history of pregnancy is unlikely to raise too much concern, but when the American retailer Target knows that a teenage girl is pregnant and starts to send her coupons for maternity items at home before her parents know about the pregnancy, there are obvious privacy concerns (Duhigg 2012a, 2012b; Hill 2012).

We have seen in this chapter a variety of approaches to data and various methods for manipulating and analyzing data. Recall that our definition of e-research involves sharing digital research materials, and this sharing, as we have seen, can take different forms: accessing a data store remotely, contributing to common resources and databases, combining existing sources of data and datasets, and sharing the work of creating and organizing data. Data are not uniform, and the uses and manipulations of data change over time, even within the context of a single long-term project.

Sharing data is currently the subject of much debate (Borgman 2015). Reproducibility is an inherent value in science, but other arguments are also frequently put forward, including the more generalized value to society of making data available for other researchers and even for new unanticipated uses. There are also increased reputational rewards that accrue to researchers who have created valuable data sets, although their work is often difficult to document or to reward in the traditional academic fashion of increased citations. And finally, there is increasing pressure from funding bodies to share databases that have been developed with public funds.

e-Research is part of a drive toward addressing questions at research fronts that require an increase in the scale and scope of knowledge production—put differently, the powerfulness of computation and data sharing. In the case of the marine biologists involved in SPLASH, it is impossible for a lone research team to answer the question "How many humpback whales live in the Pacific Ocean, and what are their habits?" Only by combining efforts has it been possible to address this basic scientific question. Likewise, for the genetics researchers trying to uncover the genetic basis for bipolar disorder or any other complex multiallelic disease, working alone or in small teams has proven to be insufficient. In any particular location, there are only so many people with a given disorder, and even if a researcher were able to recruit all of them, they would not provide a sufficient sample to enable that researcher to detect the genes responsible for increasing susceptibility to the disorder. Cooperating and sharing via computational methods and the manipulation of digital data are essential for answering these basic scientific questions. In each case, increasing the scale and speed of working with research data not only was a research requirement but also yielded unexpected scientific benefits, such as the discovery of new types of galaxies.

The styles of science that are exemplified in these cases as well as the scaling up of effort of interpreting texts (in the case of the humanities research for *Pynchon Wiki*) are enabled by research technologies and how researchers orient themselves toward a common research front. We have seen that data and data sharing have themselves become the research front for some fields of research and at various levels—within specialized research areas as well as on a national scale and beyond. The three cases covered in this chapter stand out for how they focus on distributed data sharing, but of course all the e-research in this book does this (per our definition). As we have seen especially in this chapter, bringing together data and systematizing them so as to make them manipulable constitute a particular requirement in certain areas of research, and here the focus is on the sociotechnical organization of data. We have also seen the organizational complexity of these efforts, though in cases such as *Pynchon Wiki* and Galaxy Zoo the level of complexity is relatively low. Preparing data through digitizing (although in some cases, such as social media, the data are "born digital"), cleaning, organizing, marking up, and documenting them are major tasks, but once these tasks are done, transferring, making accessible, and above all manipulating the data endlessly provide a focus for researchers and enable advances in knowledge production.

6 Digital Research across the Disciplines: The Sciences and Social Sciences

In the previous chapters, we have discussed a number of case studies of aggregating people and machines and organizing data for collaborative use. In this chapter, we pursue still more cases but use them mainly to illustrate some of the disciplinary differences and similarities in how computation and digital data are being incorporated and integrated into the practices of research. To do this, we need first to identify who the researchers are.

Who Are the Researchers?

We have so far discussed the uses of e-research without identifying who engages in e-research. Identifying both actors and excluded actors is central to the STIN strategy (discussed in chapter 2), and it is important to understand the characteristics of those who are involved with computational approaches to research. Before we discuss actors, we want to remind the reader of our discussion of a related term in the context of computing applications: *user*.[1] As discussed in chapter 2, this term is problematic insofar as it often implies the passive *use* of computers, the Internet, and other technologies rather than active selection, repurposing, moving, linking, retransmission, contribution, and creation—in addition to use—that are increasingly common both in research and among the general public. In the case of the general public, the primary value of Internet participants, or actors, is increasingly based on their active participation in the Internet ecosystem. This can be as simple as contributing to big-data analysis though the traces of one's activities, playing videogames, or interacting and sharing content with friends and colleagues on social networking sites such as Facebook or LinkedIn (Meyer, Oostveen, Schroeder, et al. 2012).

In general parlance, of course, the term *user* refers to any consumer of a product, technology, or service. Thus, a government agency making use of the results of a population survey would be a user, as would a content provider whose product is accessed across a network. *Users* here tend to be part of larger groups whose members share similar requirements and wish to achieve similar goals. By contrast, an *end user* is regarded in the technical community as any individual consumer of a product, technology, or service with needs and goals that may be distinct from any higher-order group to which they may belong. Thus, an individual citizen using an e-government portal may share with all citizen *users* the same requirement to be able to access and interact with government services but have specific characteristics as an individual *end user* in terms of which specific services they need, their individual trust levels of an online service, and their level of competence in using the service. In the main, we are interested in the social actors and participants who are typically acting as *end users* and in understanding how they regard computing tools for research, what they want from them, and the concerns they have. However, it is important to keep in mind that even when we have relied on the term *user*, one should not only think of the single-dimension of passive use but recognize that these actors are also engaged in myriad activities, some of which involve active creation and others of which are more aligned with passive use. For the individual, these shifting roles and activities come naturally, but understanding how different roles and activities should affect the design, policies, and implementation of technology is much more difficult. And here we are interested primarily in researchers, who themselves perform a variety of roles in relation to e-research, even when they may not be aware of it.

The Accidental e-Researcher

Although efforts to establish e-infrastructures or cyberinfrastructure are important, many scholars are likely to become or to have become e-researchers accidentally or unwittingly. They will not know the term *e-research*, they will not be aware that there is a generalized movement toward funding and supporting e-research, and they will be aware of collaboration tools only insofar as they are able to use them in their own research environment as a means to pursue their scholarly goals.

In our research about the social aspects of e-research, we have come across a number of cases that were not labeled as e-research or associated with e-research programs and yet seem to fit our definition. Furthermore, these uses shared features that set them apart from the other, "official" e-research projects that we were examining. For instance, many of these projects began based on developers' interests and the application domain's needs, with organic development from the ground up and features being added as the tool or resource grew. These projects ranged from small efforts to develop new tools and data sources to large distributed projects with relatively large numbers of users, many of whom were often codevelopers. Because our definition of e-research is the development of shared digital tools and data for distributed online research, the projects described in this book fit the definition of e-research even if they were not conceived as such. Indeed, several examples, such as GAIN, SPLASH (which predated "e-research"), and *Pynchon Wiki* (initiated outside of academia) can already be mentioned as accidental e-research.

e-Research has frequently been seen and analyzed from a top-down perspective. In these accounts, it is described as being initiated by research funders to develop new research infrastructures. The aim of analyzing these programs or individual projects developing within them has often been to illustrate the gap between vision and reality. In this view, e-research projects have been criticized in many private conversations as following a top-down approach or a technology "push" that does not respond to user needs. In fact, it has been suggested that much of the work in UK e-research funding programs originated from computer scientists rather than from domain experts, so there were failures of uptake and mismatches between what was developed and what domain experts needed. It has also been pointed out that top-down, large-scale, or systemic efforts often suffer from problems of conflicting standards or lack of interoperability as well as an inability to achieve sustainability in terms of resources.

The accidental involvement of researchers in e-research instead follows a pattern that is similar to what happened generally in the course of the computerization of academia. Although early adopters of computers were often quite aware of actively seeking to computerize elements of the scholarly process, mainstream and later adopters, in particular younger people who became researchers after computers had become ubiquitous in academia, are more likely to have taken for granted the involvement of computerized resources in their scholarship. Think back to the era just before computers—Did anyone spend any time debating whether the typewriter should be used instead of the quill pen? Was there a "t-research" movement, seeking to identify those for whom typewriters formed a central and invaluable tool for scholarly communication, while at the same time suggesting that research funders should advantage researchers who either employed typists and typewriter technicians or were able to acquire typing skills themselves? Of course not. Even without any systematic push toward typing-enabled research, we all nevertheless have become typists today:

most academics spend large parts of their workday typing documents on their computers, and there are no longer dedicated typists and typing pools to do this work for us. We all are accidental typists, and, similarly, we all are accidental computer technicians as we attempt to keep our sometimes fussy hardware and software functioning long enough not to crash and destroy our latest work.

In the future, it is easy to imagine academic research as having become completely dominated by e-research. In this e-research future, few will bother with trying to "problematize" the extent to which research is engaged with and enabled by electronic tools and data any more than one would spend time contemplating the impact of the invention and commoditization of paper on the academic research enterprise. Like paper, computers are starting to disappear from notice by their ubiquity. In the 1980s, when computers were new and novel, seeing a computer might evoke comment. Today, computers are no longer novel or rare (at least in the developed world) and have begun to disappear from view as items deserving special notice. If the current efforts to establish e-research are successful, the same will be true of e-research at some point in the not so distant future. We all will be e-researchers, and at the same time none of us will be e-researchers.

Disciplinary Differences in e-Research

To ground our discussion of disciplinary differences in e-research in data, we can return to the publication data discussed in chapter 3. Recall that we queried Scopus for a broad set of search terms designed to find publications from a variety of disciplines that discuss topics related to computational approaches to research. From these data, we can now look at the differences among those who are writing about e-research topics from the perspective of academic disciplines.

Figure 6.1 demonstrates some of the differences among the disciplines in this sample. The chart compares the ratio of journal publications to conference papers by discipline, and the differences are striking. According to the sample, researchers in computer science and engineering are more likely to publish their results in conference proceedings, whereas researchers in other fields and disciplines are much more likely to publish their work in journals unless they are coauthoring papers with computer scientists. If we take the social sciences as an example, we can see that in the overall sample 42 percent of social science publications are in conference proceedings—that is, 58 percent are in journals—compared to 73 percent

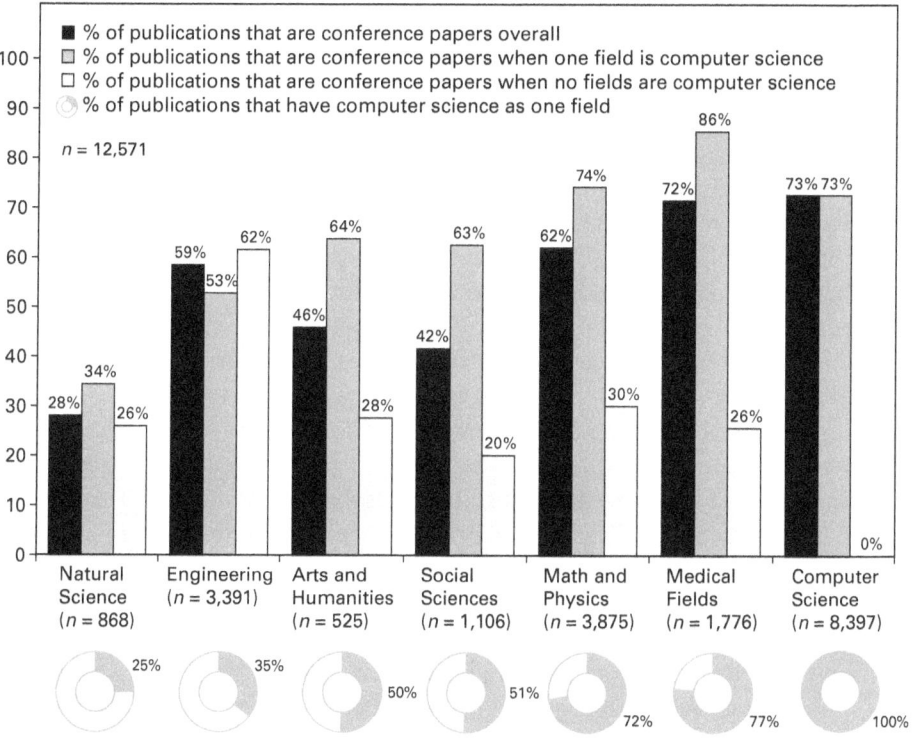

Figure 6.1
Publication type by field, showing percentage of conference papers and journal articles only, drawn from the data set described in more detail in chapter 3.
Source: Data compiled by the authors from Scopus.

of computer science publications in conference proceedings. Within the social science sample, 51 percent of the publications are jointly authored with computer scientists; considering just the joint social science/computer science publications, 63 percent of the social science publications are in conference proceedings. However, when social scientists publish *without* computer scientists as coauthors, only 20 percent of their publications referencing e-research are submitted to conference proceedings. Likewise, 72 percent of the medical papers in this sample are in conference proceedings, but when papers were authored *without* computer science involvement, the percentage in conference proceedings drops to 26 percent. This suggests that the centrality of computer science to e-research, which we have discussed throughout, also has implications for publication patterns.

Across the fields in the sample, when work is not published in collaboration with computer scientists, only 20 to 30 percent of the research is published in conference proceedings (with the exception of engineering, which has publishing norms similar to those of computer science). This is not the first time that challenges of interdisciplinary collaboration have been discussed (see, for instance, Dutton, Carusi, and Peltu 2006), but what these data underscore are the tensions that can arise between social scientists and computer scientists collaborating on publications as they decide the proper publication outlet. The conferences favored by computer scientists are not considered publications in the same way by social scientists, which can lead to differences in understanding the impact of a multidisciplinary team and lead to challenges in assessing a project's impact.

A brief anecdote from our own experience illustrates this point. During the course of one of our projects that involved collaboration among social scientists and computer scientists, the group had jointly written a paper that was submitted, accepted, and delivered at a conference and then published in the conference proceedings by one of the major computer science and engineering publishers (such as the Institute of Electrical and Electronics Engineers or Association of Computing Machinery). In the first project meeting following the conference, the team was discussing the paper, and one of the social scientists commented that because the paper had been so successful, the group "should consider publishing the paper now." The reaction from the computer scientists was puzzlement: "But we just published it!" The social scientists contended that it hadn't really been published since it was not in a journal, and conferences do not count, even when peer reviewed. The complete bafflement by players on both sides regarding the other side's position was clear and illustrates the biases brought into multidisciplinary teams because of the norms of different disciplinary homes.

These publication choices have important implications for one's academic impact, which is a topic of increasing concern across the disciplines as scholars are asked to demonstrate the impact they are having both within and outside academia. In table 6.1, we offer a comparison of citations rates by field and type of publication. We can see that in our overall Scopus-generated sample ($n = 14,064$, which includes all types of publications), about half of all publications are cited, and these cited publications are cited about six times on average. The most cited areas are medical sciences (60 percent cited an average of 7.7 times each) and the natural sciences (60 percent cited an average of 9.2 times each), which is consistent with what one would expect based on previous work on these disciplines.

Table 6.1
Citation Rates by Field and Type of Publication

	Publications (All Types)			Articles			Conference Papers		
	n	% Cited	Mean Number of Times Cited	n	% Cited	Mean Number of Times Cited	n	% Cited	Mean Number of Times Cited
Overall	14,064	49.7	5.9	4,609	70.1	10.6	7,962	40.3	2.3
Computer Science	9,123	47.6	4.4	2,295	71.0	10.9	6,102	41.5	2.0
Math and Physics	4,256	56.1	5.1	1,470	74.0	9.7	2,405	50.2	2.2
Engineering	3,774	47.2	5.8	1,407	69.7	9.3	1,984	34.8	2.9
Medical Fields	2,088	60.3	7.7	505	73.5	18.7	1,271	55.2	2.3
Social Sciences	1,256	49.0	4.3	645	64.8	6.3	461	27.3	1.5
Natural Sciences	1,059	59.9	9.2	624	73.4	11.9	244	31.2	2.2
Arts and Humanities	625	40.2	2.1	284	52.8	3.1	241	27.0	0.8

Source: Data compiled by the authors from Scopus.

The arts and humanities are the least frequently cited (40 percent), each paper cited the fewest times (2.1).

Comparing articles to conference proceedings, however, we can see a striking difference in citation frequency and number of citations. Overall, 70 percent of the journal articles in the e-research sample were cited an average of 10.6 times, whereas only 40 percent of the conference papers were cited, and they were cited only 2.3 times on average. For computer science and engineering, the majority of publications in the sample are in conference proceedings, but the citation cost of that method of publication is severe: 71 percent of computer science journal articles are cited an average of 10.9 times, whereas 42 percent of computer science conference proceedings articles are cited an average of 2.0 times each. The largest gap in likelihood of being cited is in the natural sciences (73 percent for journal articles versus 31 percent for conference papers), and the largest gap in average citations is in medicine (18.7 citations per journal article compared to only 2.3 citations per conference paper).

According to our sample, there is clearly a cost to publishing in a conference instead of in a journal, even for those fields such as computer science and engineering, which are traditionally held up as the models for fields that have chosen to focus on conferences as their "high-impact" outputs. In no field in our sample do conference papers come close to journal articles in terms of the likelihood of being cited and in the average number of citations per publication.

Another challenge of multidisciplinary collaboration in the area of e-research is that researchers in different disciplines are interested in different topics even when working together and use different languages to discuss their interests even when those interests overlap. In figure 6.2, we have generated simple word clouds to illustrate the difference in language used by three main disciplinary areas when discussing topics related to e-research. These word clouds were generated from the top-100 cited articles in computer science, the social sciences, and the arts and humanities

Figure 6.2
Comparison of top-50 words in the top-100 articles. *Note*: Illustration represents the 50 most common words among the 100 most cited titles in each data set, with common English words such as *a* and *the* removed. Data limited to single-discipline publications, using only articles and conference papers.
Source: Data compiled by the authors from Scopus; visualization made using http://wordle.net/.

The Sciences and Social Sciences

Computer Science

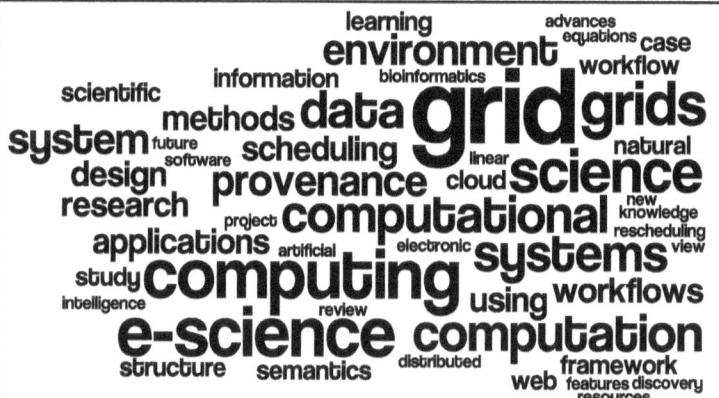

Social Science

Humanities

from our Scopus sample and visualize the 50 most frequently occurring words in the titles from each sample.

This figure helps us to see the difference in focus when comparing the words used and by extension the topics common in articles on e-research. Computer scientists have a greater focus on technical issues (using such terms as *Grid, systems, workflow, applications, environment*), whereas social scientists unsurprisingly focus on the more social aspects of e-research (using such terms as *information, knowledge, research, science*). The social scientists are also more likely to discuss problems and, somewhat interestingly, libraries. In the humanities sample, the digital humanities loom large, and topics such as history and text also make an appearance.

There are important overlaps among the three samples, such as a similar focus on data that recurs in all three, which we also identified as a common element in chapter 4. Although this overlap is more illustrative than analytical, it is useful to look at words in this way to be reminded that when researchers collaborate, they need to learn a new specialist language to understand their colleagues' interests.

Sciences

In order to understand how different disciplines are implementing e-research and engaging with the tools and methods of e-research, we can now examine case studies spread across a number of disciplines. We can start with the sciences that were among the earliest to make e-research efforts. Physics and astronomy, the two cases we look at here, were particularly active leaders in this area, developing many of the well-known tools and early-use cases for distributed computation. The first, briefly mentioned in chapter 1, is Enabling Grids for E-sciencE, or EGEE,[2] which was a European project from 2004 to 2010 ("European" because it was in part folded into the European Grid Infrastructure, or EGI, and other projects) that developed tools for analyzing data, among other things, by means of sharing computer-processing power. This project or infrastructure developed a range of other software tools that are applicable to various domains. The second case, the International Virtual Observatory Alliance (IVOA), is an ongoing collaboration (started in 2002) that is producing a tool for managing, federating, and annotating astronomical data but, again, can be used for other types of data. As we shall see, the contributions toward extending the global scope of research, reshaping disciplines, and developing research technologies are mutually reinforcing. Yet, at the same time, in each case these characteristics are *emergent*; they are instantiated, but of course there are limits to how an ongoing achievement can be captured at any given time.

EGEE and High-Energy Physics

EGEE was a European project, funded mainly by the European Union, to provide a large-scale multidisciplinary Grid infrastructure for the European Research Area (as it is known in EU research policy).[3] Although it started on a smaller scale, it grew into a large-scale collaboration that provided access to some 30,000 central processing units and several petabytes (10^{15} bytes) of storage, which were used by more than one hundred research groups from several scientific domains and from around the world. Although the EGEE project ended in 2010, the maintenance and continued growth of the infrastructure was passed to a new organization, EGI.[4] During the time EGEE was operating, it was the largest e-science collaboration worldwide in terms of scale, diversity of disciplines involved, and perhaps organizational complexity. EGEE was led by CERN in Geneva, which has conducted fundamental research in physics since 1954, focusing on experiments with large-scale particle accelerators. The most powerful particle accelerator, the LHC, is well known among scientific circles, but, somewhat unusually, also to the general public due to its role in detecting the much publicized Higgs Boson particle, a particle whose existence is critical to supporting validity of the "standard model" in particle physics. The experiments to do this have generated many petabytes of data per year that have required the use of shared computing resources across a number of sites. This kind of "big science" (Galison and Hevly 1992) collaboration, involving many institutions and hundreds of physicists, has been common in particle physics since the middle of the twentieth century (Galison 1997), but the scale of the data that need to be processed and analyzed in this case was unprecedented (Hey and Trefethen 2006).

EGEE began in 2004 as the initial two-year phase of a four-year program, which was then extended.[5] The first phase began exclusively with European members, but in its second phase (EGEE-II) it became international, with partners from 32 countries from North America, Asia, and other parts of the world. The group was dominated by partners from the developed countries of the Global North, and its geographical range reflects the fact that e-science on this scale is being carried out mainly by the countries that can afford to do so. Similarly, although the project was initially centered on the physics community (and then extended to biomedical research in the first phase), it later included many different disciplines, even though the main applications of the infrastructure remain high-energy physics and biomedicine (Gagliardi 2005).

EGEE is a distinctive e-science project because it is the largest such project, whether measured by resources (funding), number of partners, or networked computer-processing power. In fact, calling it a research "project" is somewhat odd because EGEE's longer-term aim was to be part of a European e-*infrastructure*, a goal that has now been realized. A major challenge, therefore, which is mentioned in the documents describing the project, was how to create a permanent infrastructure from short-term funding programs and funding cycles. This task entailed, according to some of the project leaders involved, finding new models for funding and for maintaining the "project" or the "infrastructures" over the long term among research policy makers at the European Commission.

The physics community's demands for high-performance computing also make this project unique. In this area of physics, there is not just a need for but a dependency on the resources that are afforded by sharing high-performance computer processing across a number of the most powerful machines in Europe and beyond. Other disciplines participating in EGEE, in contrast, do not have a central need for high-performance computing.

The relation between physics and the other participating disciplines is thus particularly complex in this case, given the range of disciplines involved. There is bound to be a two-way influence: other disciplines have had to adapt to the features required by physics as the project's core discipline, but the physics core has also needed to consider the other disciplines' needs. One example of these shared practices is the long-standing experience in physics of large-scale collaboration: multi-institutional research efforts on a large scale in physics have used memoranda of understanding for the collaborative processing and analysis of data, wherein each institution specifies the resources (funding and computing resources) it will input and what rights it has to publish the data in return. Other disciplines, however, have had less experience with predetermining exact computation needs and with this type of collaborative agreement.

EGEE took a multilevel approach to organizing the collaboration in view of the large number of participants. In terms of funding, EGEE has been a consortium with a lead partner and other partners, per the agreement with the European Commission. But there is also a federated structure based on the idea of bringing together the different regional and national Grid services and integrating them within a larger European e-infrastructure. EGEE also built on a number of existing national and regional Grids, such as the

UK e-science Grid,[6] NorduGrid for the Nordic countries, and the like. With this organization and the regular international meetings and research activities that bring the constituent projects together, it is hoped that a more permanent community can be formed around the EGEE (now EGI) infrastructure. What we can see in this case is how one community (physics) generates an infrastructure for its own needs (federating computational power in order to scale it up) by providing an infrastructure for a host of other disciplines as a means of garnering the resources for this large-scale effort. This provision of infrastructure for other disciplines, in turn, means that physics' computational approaches are extended to these other disciplines, which, so to speak, now fall under the umbrella of the physics infrastructure and adapt to these approaches.

IVOA: Federating Astronomical Data

Astronomy has been a geographically distributed and collaborative enterprise for some time because it has been necessary to use a number of telescopes in remote locations and interpret the data elsewhere.[7] Virtual observatories have come into being more recently, the term *virtual* in this case signifying that the data and images can be accessed online independently of using a telescope. The effort to bring the various national virtual observatories together into a single resource under the umbrella of the International Virtual Observatory Alliance[8] is more recent still. IVOA was formed in 2002 with the aim of developing international standards for accessing, correlating, and manipulating astronomical data. In other words, the aim was to pool data from all the national virtual observatories into a single resource and make them openly available.

The initiative began with 12 national virtual observatories (VOs) and has grown to 19 partners.[9] Membership is open to new VOs that would like to join this initiative, but they must fulfill certain criteria, which include being a major recognized national effort and being willing to abide by the procedures and standards developed by IVOA, such as allowing open access to the data.[10] The organization is governed by an executive committee and has coordination meetings several times a year. It has also established a number of working groups and special-interest groups, such as the theory group (Lemson and Colberg 2004), as well as groups to support the creation of standards and tools.

The task of developing standards has been promoted in particular by "interoperability" workshops, held twice a year, and by specialized working groups, which have concentrated on coordinating individual aspects of the larger drive toward standards (e.g., uniform content descriptors for catalog entries, image-access protocols, the VOSpace collection interface). A number of tools have been developed to support the federation of existing data repositories into a virtual repository that can be accessed from a PC.

Astronomy is interesting in terms of e-science because it involves a relatively small community of researchers (there are approximately 10,000 astronomers worldwide) with a clear and relatively well-defined task. This task requires public funding on a large scale, and it is therefore not surprising that efforts toward international coordination have taken place. Moreover, astronomy, like other science disciplines, has faced a "data deluge" (Hey and Trefethen 2003) that urgently requires common solutions. As in other disciplines, the problems in astronomy have revolved around developing common standards for data such that they can be submitted and accessed by researchers all over the world in a uniform way.

For IVOA, the task of coping with a number of sources of data has focused the discipline on creating a common interface. Describing the data with common names and identifiers has added computer science research (such as the searchability of the data via metadata) to the traditional tasks of astronomy, and these classification tools may also be adapted to other disciplines as well as being compatible with them—for example, in terms of search.

Apart from bringing data to the PC, one challenge in projects such as IVOA is how the data are to be curated.[11] The US National Virtual Observatory (which is a member of IVOA) has spearheaded initiatives to create a single virtual resource that is accessible via the Web and that brings together the data published in journals and elsewhere. In other words, there will be an end-to-end process that makes data available from capture to end-user analysis via a single VO portal with a common worldwide standard.

Social Sciences

As argued earlier, e-research needs to be given a precise definition because otherwise it might be taken to include all research using a PC and an Internet connection. Recall that our definition includes the use of shared and distributed digital tools and data for the collaborative production of knowledge. If this definition is used, it might seem at first blush that the subset of social sciences that stand to benefit from engaging with e-research is rather small. Indeed, during the course of our work studying the e–social science community in the United Kingdom, one of the early challenges was finding cases of social science where the tools of e-research were adopted in the same ways they have been adopted in the sciences. Although we found many small demonstrator projects among social scientists, we saw fewer successful efforts to build infrastructure or general-purpose tools.

Among the areas that Francine Berman and Henry Brady (2005) point to in their vision of social science cyberinfrastructure where e-research might a priori be thought to be useful in the social sciences are the role of quantitative data sets, the role of visualization, and the role of recording social interaction. These areas and many more might potentially benefit from sharing digital tools and data for jointly advancing social science knowledge across sites. But this leaves out large swathes of the social sciences, such as anthropological fieldwork, the interpretation of microinteraction in sociology, and archival work in economic history, to name just a few.

Is e–social science thus limited to quantitative social science, or will it push all the social sciences in a more quantitative direction? Qualitative social science is regarded as being more resistant to computer-enabled approaches, and yet e–social science tools have been developed for qualitative video analysis,[12] not to mention the wide variety of so-called computer-aided qualitative data analysis software programs such as NVivo, Dedoose, and Atlas.Ti. Put differently, different methods or approaches can be prioritized within e–social science. Yet the emphasis within the United Kingdom's NCeSS,[13] for example, was on certain social sciences (sociology, geography, and psychology) rather than on others (anthropology, economics, and political science).[14]

National variations have also arisen in part due to various funding programs' emphasis. For example, in the United Kingdom there has been a major focus on e–social science, though with little emphasis on business and economics, whereas in the German D-Grid initiative[15] e-research was focused almost exclusively on business applications without any social sciences represented. In Germany, despite an ambitious e-research program that includes many natural science and humanities projects, the uptake in the social sciences has so far been conspicuous by its absence (Schroeder, den Besten, and Fry 2007). In Sweden, although there was initially a plan to develop e-infrastructure to support a wide range of social science research, the plan was subsequently more narrowly focused on creating a facility for sharing microdata (Axelsson and Schroeder 2009). In the European Union, within the European Strategy Forum on Research Infrastructures, the main effort has been to develop infrastructures around existing large-scale quantitative data sets, such as the European Social Science Data Archives and the European Social Survey. These kinds of efforts create commonalities across national programs, such as when national social science data sets are being federated via the Council of European Social Science Data Archives and similar organizations around the world. These national research programs and funding initiatives will, at a minimum, shape the kinds of infrastructures and communities of researchers that develop in e-research and how e-research is used in the social sciences. And, again, we have noted a recent shift in the social sciences toward big data in the analysis of social media and the social sciences more widely, which often falls within our definition of e-research (Schroeder 2014). Thus, although we have focused in this section on infrastructure efforts and e–social science programs, these efforts and programs are a moving target as the research front moves in new directions.

VOSON: A Tool for Studying Online Networks

The building of these large-scale infrastructures is ongoing.[16] Yet there are also Web-based tools that create smaller "infrastructures" for research. Here we can take as an example the Virtual Observatory for the Study of Online Networks project,[17] which has created a tool for doing social science research on online networks, though this tool, as we shall see, can be applied to many domains. Based at the Australian National University, VOSON is a research project to study online networks. This effort had been ongoing for several years under the leadership of Rob Ackland, but it began as a formal project in 2005 when it was funded by the Australian Research Council as part of a special research initiative for e-science support. To date, VOSON activities have focused on the development of new social science research methods and tools and especially "webometric" approaches that use data from the Web such as the hyperlinks between web pages to identify, among other things, the visibility of sites (Thelwall 2009). One goal of the VOSON project is to produce software to enable these new research methods and develop tools that can be shared and used collaboratively by researchers. In this sense, VOSON is a good example of e–social science: using e-research technologies to enable new forms of collaborative social science research.

A number of features of this project are noteworthy. First, as just mentioned, the main thrust of the research is to build tools and data that can be used in collaboration among all researchers. The project has deliberately adopted an open-source approach to software development (we discuss these "open" approaches further in chapter 8), with the aim being to make the tools and data available to other researchers and encouraging them to do the same. The project has also adopted a Creative Commons approach to licensing and follows various open standards (such as for metadata).

Second, VOSON is part of a small community of researchers using certain types of novel tools (webometrics) in a relatively new area of research, so there has been a highly focused research effort with a great deal of interchange between researchers. For example, VOSON has worked with several institutions worldwide that do research in similar areas and with which collaborative research has been undertaken, including in the United Kingdom, the United States, and other parts of the world. This global nature is reflected in how one can use the tool: either as a web application or as a plug-in to the widely used NodeXL tool.[18] NodeXL is an e-social science story in itself. Originally developed by Marc Smith and colleagues at Microsoft Research, it has since spun out into a software platform supported by the Social Media Research Foundation.[19] It is a free and open-source template for Microsoft Excel rather than a stand-alone application, which makes it very lightweight to install and more accessible to a wider range of users than some of the more specialist tools for doing social network analysis. Recall here our argument about research technologies, how they travel in and out of domains and how they can be combined, recombined, extended, and continually refined and enhanced to enable the more powerful manipulation of data.

In view of the relatively small number of groups doing work on VOSON (and on NodeXL) and the fact that all the researchers have to keep in contact to remain abreast of rapidly moving developments, it is appropriate to speak of a research front here as well as of a movement or a research community to promote a particular approach to research. This community is reinforced by the fact that the researchers share a common object (here, web links) and a common approach to social science, which falls under the SNA umbrella—a larger research community that also extends beyond social science disciplines (including physics [Freeman 2004]). And although webometrics as an area within e-social science is relatively new, at the same time there has been a rapid increase in the number of papers and scholarly interest in this field.

This research area also poses a number of challenges: VOSON and similar research projects measure online networks and thus the visibility, for example, of activists and parties, but they do so for the online world. As with webometrics generally, one question concerning these online networks is their significance for offline relationships. If, for example, certain environmentalist groups are more closely linked online with other such groups, what are the implications (if any) for how environmentalist groups work offline (Ackland, O'Neil, Bimber, et al. 2006)?[20] The solutions to these challenges will require integration with and testing against other social science findings.

VOSON illustrates the global scope of social science knowledge insofar as its analysis of online networks covers the entire (global) web network as its object of research. The limitation in this case comes back to the challenge mentioned a moment ago: Even if the scope of social science in this area is global, how well does the global Web map onto other globalizing processes apart from these online networks? As for disciplinarity, one question is, Which discipline does this project "belong" to? This question poses a problem for the dissemination of the work: Should the work be presented and published in traditional political science or sociology journals (including specialisms within them, such as social network analysis), or should the publications focus more on tool development and webometrics (with all the problems of "credit" given for this type of work), and should they target journals in computer science or information science? The VOSON project has so far taken the direction of conducting social science research using a variety of methods from other fields, and yet the output has been targeted primarily at social science journals and conference proceedings. However, this approach has met with mixed success: novel methods and insights are met with enthusiasm, but there is also a lack of understanding about these new methods and about the relevance of web data to traditional social science concerns.

One way to get around the lack of understanding of this new area is, of course, subdisciplinary specialization. Work on webometrics is well suited to new outlets such as conferences and journals for e-science. It can be foreseen that this area, like other such areas, will not only become one small domain within other areas, but also form a subdiscipline or specialism of its own. One indication of this specialization is the appearance of specialist journals devoted to the topic, most notably *Cybermetrics*, founded in 1997.[21] It is also noteworthy that a number of disciplines are involved and that physicists, computer scientists, library and information scientists, as well as social scientists have been a strong presence in this type of analysis. VOSON fits this multidisciplinarity, with several computer scientists involved, whose papers often contain detailed accounts of the advantages and disadvantages of different types of software (Ackland 2009), but also papers written by political scientists and researchers from other social science disciplines such as media studies (Ackland, O'Neil, Bimber, et al., 2006). This diversity illustrates the key point that Terry Shinn and Bernward Joerges (2002) make about research technologies: that knowledge is transferred via the skills needed in the development of generic devices that need to be applicable in a range of settings. However, as mentioned in the brief discussion of VOSON and changing research practices in chapter 1, an additional challenge related to research technologies is that they must be maintained and updated to keep the momentum of this movement or research community among its users going.

Google and Big Social Science Data

Webometrics allows social scientists to see the web as a giant interconnected set of links, the patterns of which can be identified, thus creating maps of the world's knowledge and information.[22] Another area that has become a major focus of research is the analysis of search engine behavior, which offers social scientists a look—at least partially—at what is on people's minds when they are looking for information online. Here we can single out one example (Waller 2011), though a number of others (Hindman 2009; Tancer 2009; Segev and Ahituv 2010) might be used as well. Vivienne Waller had access to "transaction logs to provide an analysis of the type and topic of search queries entered into the search engine Google (Australia) in April 2009," with Google having an almost 90 percent market share in Australia (2011, 761). She also had data from the marketing company Hitwise Experian about which of 11 lifestyle groups—broadly comparable to socioeconomic stratification groups—searched for which search terms. She analyzed almost 1 percent of all search terms for a month, extracting a sample of 60,000 search terms, which accounted for 28.7 percent of all search queries (a query typically consists of two or three terms). She then used 78 codes and amalgamated these codes into 15 broad subject groupings, such as "high culture" and "popular culture," "e-commerce," "weather/time/public transport," and the like.

Waller found that "queries about popular culture and e-commerce account for almost half of all search engine queries" and, "somewhat surprisingly, [that] the distribution of topics of search query did not vary significantly across different Lifestyle groups for the broad subjects of popular culture, e-commerce, cultural practice and adult" (2011, 767). This finding is quite surprising because it might be expected that different lifestyle groups or demographics or expert versus skilled Internet users would search for different things (for examples, see Tancer 2009 and Dutton and Blank 2011). Yet it seems that, in Australia at least, users from different socioeconomic groups have similar queries. Waller found other interesting things—for example, that people looking for information "on particular contemporary issues accounted for less than 1 percent of all search queries. Queries about government, including programs, and policies, accounted for less than 2 percent of all Web search queries" (2011, 769). Overall, she argues that search engines are mainly a technology for leisure or consumption and less one for seeking knowledge and information.

Waller's analysis is a good example of an area of research we mentioned earlier, "big data." But in view of the hype about this concept, we must ask: What's new about big data in the social sciences? After all, haven't the social sciences been engaged in big data for a long time? Perhaps, yes, if we think of censuses and the like. Yet here we argue that "big" data can be defined as data unprecedented in scale and scope in relation to the object or phenomenon under investigation (again, recall our realist definition of science and our definition of data in chapter 5). Waller used data on a whole month's worth of all searches for a population, which, because of Google's near monopoly share in Australia, can be seen as an unequalled and powerful sample (although even in this case, she used a sample of the entire data set rather than the whole of the data set, which might have been a more purely big-data approach). Nevertheless, the "object" here is larger in scale and scope in terms of access to what kind of information people are seeking. Bill Tancer (2009), who is head of research at Hitwise Experian, has much bigger samples (though they are proprietary), including about 10 million Americans for many years' worth of searches, but Waller's paper is published in the top peer-reviewed journal in information science. In any event, if big data is defined as advancing knowledge with a data set of a size and scope that is a magnitude larger than any previously available within the domain, then her data count as big data.

It might be possible, of course, to obtain a nationally representative sample for Australian search queries by other means, such as a survey of users (asking them to record their search queries) or logs from the computers of a large sample of users over the course of a month or more. But it is interesting to imagine the resources required to use either of these methods, which are likely to be prohibitive for academic social scientists. It is also interesting to reflect on the research ethics constraints involved in this research, which might also present barriers. Instead, Waller's paper is a case of using data that was not intended for social science research but that nevertheless can be employed on an unprecedented scale to advance social scientific knowledge about what particular groups of search engine users search for.

There are applied or pragmatic reasons for research on search engine behavior: figuring out how to enhance the users' experience when their attention is directed to targeted advertising and ultimately predicting what they are interested in buying. But Waller has also derived powerful social science insights from these data, which include, as mentioned, that most searches are for leisure and that search engine behavior is generally similar across lifestyle groups. One problem highlighted by Waller's study (and by other studies that use similar methods) is that it is based on proprietary data. This means that even if these data were made accessible to other researchers, we still would not know how the search engines work (this, after all, is Google's secret) and thus how the data are arrived at. A second problem is that the classification of groups of information seekers is based on marketing company categories ("lifestyle groups"): these categories are similar to but not equivalent to the standard categories used by social scientists. Commercial data also raise another question: Which part of the population does not use Google, and does this differentiation give rise to the issue of representativeness? Put differently, who is being left out when the Australian population is being analyzed as "Google users"? In short, though such research provides new and very powerful insights, it is not replicable and cannot be built upon with standard classifications. It can only be built upon inasmuch as others, inspired by Waller's striking findings, try to see if they can obtain similar results. Thus, a number of issues are attached to this type of research, some of which (What does this type of analysis about the "real world" tell us about the needs of information seekers?) are similar to those mentioned in connection with VOSON. As powerful as this kind of analysis is in contributing to social science knowledge, it must nevertheless overcome these issues.

Mike Savage and Roger Burrows (2007, 2009) raise the broader concern whether private companies with this type of data are able to do more powerful research than academic social scientists. In the case of search behavior, a legitimate question might rather be whether this advance in knowledge is so powerful that it can predict people's behavior and thus enable manipulation of them: a concern in the commercial world and for privacy and autonomy, but hardly one that academic social scientists have had to worry about in the past. Similar concerns and issues about the validity and replicability of research have been raised in a number of big-data studies in the social sciences, a rapidly growing area focused on Twitter, Facebook, *Wikipedia*, and similar digital objects of research (Schroeder 2013b). The benefits of being able to analyze people's online relationships on such a large scale and with ready-made sources of digital data and powerful tools, the drawbacks of the representativeness of samples and lack of privacy, and ultimately the manipulability of human behavior—these are issues that the social sciences will need to address for some time to come.

Conclusion

In this chapter, we have begun to discuss how disciplinary differences are enacted with relation to digital research, with a particular focus on the sciences and social sciences. In the next chapter, we continue this exploration of disciplinarity, but with a focus on the humanities and on accessing knowledge. At the end of chapter 7, we revisit the cases raised in both chapters to discuss how the machinery of research is more visible and measurable and how these changes are unevenly distributed across the disciplines.

7 Digital Research across the Disciplines: Humanities and Access to Knowledge

The digital humanities as specialist sets of tools and techniques have been around for more than twenty years but became much more prominent around the middle of the first decade of the twenty-first century.[1] Since then, there has been growing interest and activity in creating digital humanities projects, in attendance at digital humanities conferences, and in the number of papers and sessions devoted to digital concerns at mainstream humanities conferences.[2] In the United Kingdom, the funding agency JISC invested heavily in the digitization of humanities content, funding dozens of projects of varying size since 2004, as has the Arts and Humanities Research Council. The US National Endowment for the Humanities, Canada's Social Sciences and Humanities Research Council, the European Union, and a wide variety of public and private funders around the world have been committing money to digitizing the historical documents, cultural materials, and other resources that make up our shared cultural heritage.

Humanities scholars have by and large responded enthusiastically to the availability of digital resources. Consider, for instance, the following reactions from researchers engaging with two popular digital resources:

British History Online is my favourite and first source for primary sources in British history. As a student of history, librarian, and writer, I return again and again. Even when I'm not researching, I often visit BHO for the sheer fun of what I might learn and discover. The site is easy to navigate, convenient, and its offerings thorough and accessible. Where else online can I find such a bounty of Britain's heritage? It is a generous endeavour and an absolute goldmine. (qtd. in Blaney and Webster 2010, 7, cited in Meyer 2011, 2)

I'm not joking but [the University of Oxford podcast site] has become my favourite site in ten seconds flat—can't stop downloading! Where has this been all my life?????? This is ridiculous! (qtd. in Wilson, Marshall, and Geng 2010, 7, cited in Meyer 2011, 2)

These admittedly hyperbolic reactions are nevertheless consistent with our data about how humanities scholars view the digital resources available to them. In a survey of 426 humanities scholars in 2009, 83 percent of respondents described themselves as either enthusiasts or advocates of digitization in the humanities, and only 3 percent counted themselves as either critics or skeptics (Meyer, Eccles, Thelwall, et al. 2009, 151). In the same study, 98 percent of respondents considered digitized collections to be useful, and 76 percent felt that new questions would require the use of digitized resources. Conversely, only 2 percent felt that the digital approaches threatened to undermine the quality of research in the humanities, and only 5 percent felt that digitization was more hype than reality (152).

Compare this level of enthusiasm to that shown in the social sciences for the tools of e-research, and the popularity is clear: in a survey of 526 social scientists in 2008, only 60 percent considered the digital tools of e-research to be useful, and 59 percent felt that there would be new questions in the social sciences that require the use of digital tools and methods to answer (Dutton and Meyer 2009, 231). Likewise, although there were, as in the humanities, few outright skeptics of digital research in the social sciences (8 percent), only 33 percent considered themselves enthusiasts or advocates (compare this low percentage to the 83 percent of humanities scholars enthusiastic about working with digital materials). The largest proportion (43 percent) of social scientists considered themselves merely observers of the trend toward integrating shared and distributed digital tools and data in their research (Dutton and Meyer 2009, 230), even though the sample for the cited survey was skewed in favor of those who had shown interest in e-research projects and programs in the past (Meyer and Dutton 2009, 241).

One explanation for the difference between social scientists and humanities scholars is that the digital research materials in the humanities, particularly with regard to the initial digitization efforts, are familiar. If a medieval musicologist, for instance, consults physical manuscripts in an archive, they might use tools such as "a magnifying glass to inspect details and a ruler to measure its size. They also inspect physical characteristics such as leaf textures, bookbinding, and ink pigments," among many other characteristics of the manuscripts (Eden, Jirotka, and Meyer 2012, 73). When using digitized collections, such as the Digital Image Archive of Medieval Music,[3] the interfaces provide a digital equivalent of familiar tools: "each image includes the following elements: a manuscript page or fragment, a colour patch and ruler. Both the colour patch and the ruler are placed alongside the manuscript item serving as reference points" (Eden, Jirotka, and Meyer 2012, 73).

Likewise, consulting digitally scanned copies of newspapers, books, and other printed materials is not an unfamiliar experience for the humanities scholar accustomed to sitting in the proverbial dusty archive. The scanned copies are often high-quality reproductions of the material artifact, just as readable as the original, if not more so due to the ability to zoom and resize the image on the screen. The digital copies also have distinct advantages over their paper counterparts from the researcher's point of view. First, there is no inherent barrier to creating digital materials, and they can be accessed anytime, anywhere (within the limitations set by institutional subscriptions for nonfree resources); travel is not required; multiple sources can be consulted simultaneously; the cost of taking a copy (within the limits of copyright) is essentially nil; and additional operations can be performed on the materials (if the affordances of the technology allow it).

Digitized Humanities Collections: Impact on Teaching

One of the ways that digital resources are increasing their impact in the humanities in the long term is by becoming embedded in the work practices of new scholars who are just learning their craft.[4] On the one hand, the so-called digital natives—the technically savvy students currently in undergraduate and postgraduate programs—are less likely to be wary of digital resources and may be more likely to embrace them (although these students' actual abilities is less clear in practice; see Helsper and Eynon 2010). For instance, when the Siobhan Davies RePlay project[5] asked students being trained to use RePlay to share videos of dance practice and reflections about dance, almost all the students were already using Facebook and YouTube to post thoughts, comments, images, and videos; the challenge, then, was to get them to think about RePlay in the same terms, as a place to document their dance process in a way that is as immediate, relevant, and easy as documenting their social lives on Facebook. In another example from the performance arts, one of us (Meyer) and Isis Hjorth observed teenaged dancers and musicians collaborating to create a new performance:

> The eight young performers in Forbidden Rhythm ... used their mobiles in an interesting collaborative way, however. When the group was working on a new portion of the show, the beatboxers were sometimes tasked with each creating a one or two bar piece of beatboxing, which the dancers would then create choreography to. The beatboxers recorded their segments on their mobiles, and then sent them to the

dancers, who would listen to the piece and create matching dance moves, and then rehearse them by listening to the clip on a loop. Not only did this allow for rapid creation of collaborative work (since multiple performers could be creating music and dance at the same time in parallel), but it also resolved disagreements later in the process. For instance, on one occasion, the beatboxers changed a rhythm slightly, which threw the dancers' choreography off. A quick consultation of the mobile phone recording got the rehearsal quickly back on track. (Meyer and Hjorth 2013, 24)

The impacts of digital materials on teaching can transcend disciplinary boundaries once the materials are easily available online. Old Bailey Proceedings Online (OBPO),[6] for instance, found evidence that although the OBPO was used by courses focusing on crime and criminal justice, it was also being used by courses focused on the use of primary sources, such as special subjects, and methods-and-skills courses.

Another digital resource, the Oxford University podcast collection,[7] also draws in students because of the easy and free availability online at its own website and via iTunes U but then keeps the students there because of the quality and relevance of the podcasts. Comments on the "excitement and clarity" and "exemplary and passionate delivery" demonstrated in the podcasts reflect students' enthusiasm for the material (Wilson, Marshall, and Geng 2010, 9).

Features of a collection can also draw teachers to include the resource because it fulfills particular pedagogical needs. The OBPO, for instance, reports that "some teachers use the statistics function to introduce students to quantification, noting that it provides an 'unscary' way of introducing students to numerical analysis" (qtd. in Howard, Hitchcock, and Shoemaker 2011, 4). Histpop[8] interviewees pointed out the same effect:

A lecturer in Historical Geography at King's College London reported that he had used Histpop in both second and third year undergraduate courses in Urban Historical Geography, allowing students access and freedom to explore primary sources at this stage of their undergraduate careers. This, he said, led to several Final Year dissertations of high quality, which he said were enhanced by early access to primary sources. "I think [the students scored so well] ... because they had already been used to dealing with historical documents ... using Histpop as a teaching tool fed into those particular dissertations." (qtd. in Meyer, Eccles, Thelwall, et al. 2009, 26)

RePlay also fulfills an interesting niche for the teaching of dance by providing access to dance works otherwise unavailable to teachers:

[RePlay] helped me to develop new teaching methods. Makes teaching more interactive. Students feel much more engaged with professional dance world when they analyse work from SDDC RePlay. Access to older works e.g. Sphinx is vital to delivery of A Level Dance spec. as it's difficult to get hold of dance works. I am able to do a successful scheme of work based on Siobhan Davies, whereas with Alston and Cohan there aren't many of their early works available, so I talk about their early works but have no clips to show the pupils—therefore they don't learn as well. (qtd. in Marsh and Evans 2010, 12)

Other projects also served to support teaching. A Vision of Britain,[9] for instance, reported this user feedback:

I want to congratulate you on the splendid work you've done. This semester I'm teaching a course on the literature of the British countryside, and imagine my delight to find how easily my students can locate on-line excerpts from Defoe's *Tour of Great Britain*, Young's *Annals of Agriculture*, and Cobbett's *Rural Rides*. Without your work, this course would be much poorer in content.—U.S. University User, January 2009. (qtd. in Aucott, Healey, and Southall 2011, 3)

These sorts of small-scale impacts disappear when comparing broad numerical measures (a site such as RePlay, for instance, receives far less web traffic than more widely used resources), yet they indicate that when the students have access to a resource, the impact can be significant even in small numbers. One user, for instance, indicated that it is RePlay's easy availability that resulted in the decision to add Siobhan Davies to the curriculum.[10]

The Siobhan Davies Replay project had a central pedagogical focus, which was to embed RePlay into students' personal development plans. To support this focus, students took part in a series of seminars and workshops focusing on both the dance content and the skills needed to use RePlay as part of their development:

The introductory session provided students with a range of "guided tours" through the archive to identify content via different enquiries. Students then created their own virtual "scrapbooks" by browsing, searching and finding content on RePlay. These "guided tours" are posted on the project blog and will be added to RePlay to assist other users to navigate through the archive. (Whatley, Barzey, Marsh, et al. 2011, 7)

In this way, the students became both users of and contributors to the RePlay resource.

Teachers also have a desire to be able to share their work practices and examples with their students. One OBPO user who was enthusiastic about the new workspace features that OBPO introduced because he or she could use them to teach students how projects can endure over the course of time and as a way to share the searches in the teacher's workspace as a way to provide examples to students. Here we can see research technologies being developed as collaborative workspaces.

Teachers using the University of Oxford podcast resource commented that the enthusiasm and clarity in the podcasts were an inspiration for their own teaching and gave them new ideas for material and for methods of presentation. This type of impact is very indirect and difficult to measure beyond anecdote, but it nevertheless suggests that collections that can stand as exemplars have the possibility of spreading their influence indirectly by inspiring similar levels of quality in other resources.

The opportunity for students to engage with primary source material at a much earlier stage in their academic careers has been discussed elsewhere (Meyer, Eccles, Thelwall, et al. 2009, 26, 106, 148). Students in focus groups reported this aspect of working with OBPO: "For all of them, it had been their first real opportunity to engage with a substantial set of primary sources that had not been pre-selected by a teacher. This was something they found exciting and stimulating because it allowed them to find and interpret material for themselves" (Howard, Hitchcock, and Shoemaker 2010, 7). Histpop was similarly noted for its impact on students: "Histpop [makes] it possible to do a completely different project [at the undergraduate level]. ... It allows them to start using primary sources and do some basic research, which otherwise they wouldn't be able to do" (qtd. in Meyer, Eccles, Thelwall, et al. 2009, 26).

Digital collections' role in engaging next-generation researchers is not unique to the humanities. In our teaching of social science students at our own institution, for instance, the ready availability of data from the Internet clearly encourages students to make use of the possibilities of research using digital methods. In natural sciences, too, myriad online tools and resources, many of which fall within our definition of e-research, have become available in teaching students at various levels. How they will be reflected in the research practices of the next generation of academic researchers remains to be seen.

Large-Scale Textual Analysis

Our next case focuses on the analysis of text from collections of books, a practice that has made incursions into history and literary studies.[11] In discussing humanities in this chapter so far, we have focused on teaching and research practices; in discussing textual analysis, however, we can ask whether this type of research also contributes novel questions or insights in well-established domains or disciplines. We could have also asked this same question for the other areas and disciplines that we have discussed in this and previous chapters, but this is more difficult in some domains in which it is less easy to identify research fronts. The study we focus here on also explicitly claims that new questions are being enabled. This "culturomics" study utilized the digitization of more than 5 million books (approximately 4 percent of all books ever printed) through the Google Books project (Michel, Shen, Aiden, et al. 2011), examining a corpus of more than 500 billion words, which easily fits our definition of big data (data a magnitude larger than any previous research in the domain) and represents a new form of interaction with research materials in the humanities.

In this case, the study produced novel insights about language use: for example, charting the rise and fall of the English term *feminism* and the French term *feminisme* over the course of the twentieth century provides clues about the cultural significance of this phenomenon in different cultural contexts or periods. This is just one of several examples in the study. More generally, the study claims that this is a "new science"—hence the suffix *-omics* in *culturomics*—and that "culturomic results are a new type of evidence in the humanities" (Michel, Shen, Aiden, et al. 2011, 181). The study encountered criticisms (described later), but because it was not undertaken by humanities scholars, we can add another example of the use of "quantitative methods" in the humanities: the Litlab at Stanford University analyzed word frequencies in 2,779 nineteenth-century British novels (from a commercial database of digitized books) and found that over the course of the century there was a decline in "abstract values words" and a rise in "concrete, physical, specific, and non-evaluative" words, which Litlab called the "hard seed" (Heuser and Le-Khac 2011, 83). This study, again, claimed new findings: the authors hypothesized that this shift in word usage could be linked to social change, in particular rapid urbanization, in British society (Heuser and Le-Khac 2011, 85).

Whereas "culturomics" promotes a scientific approach to the humanities, quantitative methods, as used by the Litlab, has championed a new approach within literary studies: "distant reading" (Moretti 2000, 2005, 2011), in contrast with the "close reading" of highly interpretive approaches to literature. The "distant reading" and "culturomics" studies do not have the replicability problem of Vivienne Waller's (2011) study of search behavior described in the previous chapter; in fact, one of the promising aspects of this research is that others can add to and extend these studies: comparing their results with the results of examinations of other bodies of texts, comparing the results for other key words, and the like. Indeed, unlike the

Litlab study, which requires access to a commercial database of literary texts, Google Books' Ngram Viewer allows anyone to search this corpus for word frequencies. Thus, whatever one may think of these sorts of studies, they hold the promise of a rigorous, quantitative, hypothesis-driven, and systematic approach to the study of culture and literature using patterns of words on a large scale.

Yet these studies have also drawn criticisms. For example, Andrew Stauffer argues that Ryan Heuser and Long Le Khac's (2011) link between word frequencies in novels and a shift in British culture in general is a large leap for which more evidence and interpretation are needed (2011, 65). Questions have also been raised about the quality of the data in Google Books (Duguid 2007; Nunberg 2009). Perhaps the main concern here has been the scientistic challenge to cultural and literary interpretation, which is a question of disciplinary turf. This is why "distant reading" has provoked more reaction than "culturomics": because the challenge has come from within the study of literature and published in humanities journals, it has ruffled feathers among other literary scholars (e.g., Fish 2012), whereas scientists publishing on culturomics in the journal *Science* perhaps seem less threatening.

In the case of textual analysis, the concern should not be with the purity of literary studies or the superficiality of word counts for cultural analysis. Instead, we should ask how quantitative approaches complement and add to more interpretive ones: the latter will not be displaced because literary and cultural studies will continue to be pluralistic (low task certainty and low mutual dependence), but they will need to make peace with complementary quantitative approaches, which will continue to be taken because they are cumulative and constitute a distinctive niche within literary studies. Put differently, even if these quantitative approaches contribute new questions and insights, they need to become integrated within the humanities disciplines (history, literary studies, and others) of which they are a part.

Digital as a Dirty Word

It is not just approaches such as culturomics and distant reading that cause disquiet among some scholars in the humanities. Indeed, for many, the very notion of digital research in the humanities still has a touch of illegitimacy, even in the case where the scholars are enthusiastic users of digital tools and digitized collections. In an interview for our research, one humanities scholar told us, "I do feel pressure to work more with originals than with the digital images, but for the most part I do feel like I get more out of using these images on my computer. But there's a certain pressure that that's not what top scholars do because that's not what top scholars did 25 years ago" (interviewed by project staff, June 14, 2010, Cambridge, Massachusetts, by telephone).

This sense that serious humanities scholars must somehow hide their use of digital resources or pretend not to be enamored of digital research is a common theme in our work with humanities scholars, even when, as reported earlier, more than eight out of ten humanities scholars say they are enthusiastic about digital collections, and 98 percent find them to be useful (Meyer, Eccles, Thelwall, et al. 2009). These biases—or predilections, depending on one's viewpoint—are learned very early in a scholar's career: during a focus group made up of second-year history undergraduates, the students discussed digital resources. They all agreed that sources such as *Wikipedia* should not be cited, and they thought that their professors forbid them from even consulting *Wikipedia* (although they all admitted to doing so anyway, indicating that it was their dirty secret that they found *Wikipedia* helpful, particularly in identifying more mainstream published resources to consult). Furthermore, the students reported rarely visiting libraries and consulting printed materials, instead preferring sources that they could consult on their laptops through the library's electronic resources. However, with respect to citing sources, a number of the students reported that they would look at their overall bibliography, and if it appeared to have "too many URLs [Uniform Resource Locators, unique addresses for items on the Internet]" (which they agreed was probably more than one-fourth to one-third of all their citations), they would simply start deleting URLs until they appeared to have a balance of sources that leaned in favor of printed materials.

This worry about citing digital materials is a regular theme in our work with the humanities community. In a 2008 survey we undertook of humanities scholars, we found that users of four electronic resources would cite only the original version of a publication, with no indication of the digital source, between 36 percent (for 18th Century British Official Parliamentary Publications) and 53 percent (for British Library Newspapers) of the time. In a fifth project in the study, more users reported giving either a URL or a combination of the printed version citation and a URL in citations (91 percent for Histpop), but this greater percentage was due in part to the fact that Histpop provided a suggested citation that included a short URL (Meyer, Eccles, Thelwall, et al. 2009, 147). In a more recent study, we found that users of EEBO-TCP texts were still in a third of cases likely to cite the paper versions of books (34 percent), even if they had never seen them (Siefring and Meyer 2013, 17). In a case study of *British History Online*, the authors noted: "One of the history teachers we interviewed commented that he insisted upon his students citing print versions of text rather than the online version. His reasoning was that a string of arbitrary numbers is a very unhelpful form of citation because it tells the reader nothing about the source being cited" (Blaney and Webster 2010, 9, qtd. in Meyer 2011, 40–41). The *British History Online* collection's approach to addressing this particular criticism of online links is worth noting: the adoption of Cool URIs (Uniform Resource Identifiers), in which human-readable meaning is embedded.[12] For instance, the URL for the *Journal of the House of Lords* on *British History Online* now follows the format www.british-history.ac.uk/lordsjrnl*n*, where *n* is the journal number. Unlike the confusing and sometimes excessively long links often generated by database-driven web applications, these links give the human reader contextual information about the link *without having to visit the link itself*, just as seeing a citation to a work provides context to the knowledgeable reader without requiring that he or she consult the original document (Blaney and Webster 2010).

Of course, there is one well-known problem with the current Web: the problem of disappearing links, also called "linkrot." According to one study, "13% of Internet references in scholarly articles were inactive after only 27 months. Another problem is that cited Web pages may change, so that readers see something different than what the citing author saw" (Dellavalle, Hester, Heilig, et al. 2003, 787, qtd. in WebCite 2010). Beyond the inconvenience, an insidious unanticipated consequence of linkrot is that it reinforces researchers' habit of including as few URLs as possible in their scholarly work. Thus, when online resources try to assess their impact using techniques such as webometrics, the lack of cited links makes their resource appear to have less of an impact (recall our argument about research "visibility" in chapter 1). Second, it makes readers trying to track down the sources of information work much harder as they try to discover not only the correct source of the citation, but also the version of the source that was cited because the pages may have changed considerably. This problem also reinforces the notion that *digital* is a dirty word when it comes to humanities research.

We do not want to portray the humanities as antitechnology, for generally we have found the opposite: an eager embrace of digital tools and materials. However, one of the disciplinary biases of the humanities is the persistent sense that research should not become too reliant on digital information, which is in stark contrast to view supported by physical scientists, who report not only happily citing URLs but also rarely printing or even saving copies of articles that are stored in databases online. The physical scientists told us that they no longer felt the need to store and organize the articles themselves because they could just go back to the online copy and consult it whenever necessary (Meyer, Bulger, Kyriakidou-Zacharoudiou, et al. 2012).

Conservatism is not unique to humanities scholars, however. Here we can give Alyssa Goodman's example of the introduction of a new tool in astrophysics (provided in a recent talk;[13] see also Goodman and Lintott 2013). Goodman, of the Harvard-Smithsonian Center for Astrophysics, discussed a new version of the Astrophysics Data Service literature search system called "ADS Labs,"[14] with a more modern interface than the original version as well as many new features, including the ability to link through to data sets and to the Worldwide Telescope[15] directly from references to astronomical objects. However, even when the system had been available for several years, there was little uptake in the astronomy community, with only between 2 and 20 percent of the community using it (Goodman and her colleagues had different estimates). There was considerable frustration that the community had not more actively embraced the new platform, a problem that could be attributed mainly to inertia or the routine of working with the existing interface. In sum, what we can see in the uses of digital tools and resources in research, as in teaching, is an ongoing shift toward the online world, but a shift is difficult to capture because it is so varied and—again—ongoing: science (or knowledge) and technology in the making rather than already made.

Digital Transformations of Desk Research

In this chapter, we have discussed both knowledge production and new research practices, including for teaching.[16] An overlooked area in this account is how research is accessed more broadly considering that desk research, such as consulting secondary research materials, is still an important type of research across all disciplines. In the humanities, consulting both primary and secondary materials constitutes a major type of activity for researchers, and, as discussed earlier, the humanities have been particularly transformed in terms of how desk research is done.

In recent years, there has also been a wider transformation in how formal and informal science communication is disseminated by electronic means (Nentwich and König 2012). Although we still know little about how this transformation has affected the nature of research, particularly in light of disciplinary differences, there is no reason why the simple growth and proliferation of outputs should lead *straightforwardly* to a richer and more diverse information and knowledge environment. Instead, gatekeepers such as search engines shape online visibility, which, combined with competition for limited attention space at the leading edge of research (there are only so many research fronts that, for reasons of specialization and limited time, a researcher can keep track of), leads to a different model of how access to knowledge and information is being shaped.

To make this argument in more detail, we can ask how researchers gain access to knowledge at a time when scholarly communication and materials are increasingly moving online. This topic has so far been discussed mainly in terms of journal publication and readership. Here we take a broader view, in part because the focus on journals overlooks a number of trends favoring e-research, where knowledge production and dissemination are broader than journal publications: e-research also consists of efforts to develop distributed online tools, data, and other resources, as we have seen throughout the book so far. Another reason to take a broader view extends the horizon still further: the changes in scientific communication and collaboration are happening beyond the research community as well. For example, search engines affect what can be found online generally, among broader publics. New search behaviors are particularly evident among a new generation of scholars and researchers, and thus a wider picture is needed because search results are, in turn, affected by search behavior.

It is worthwhile to think of these changes in terms of a schematic representation of the overall scholarly communication ecosystem. Figure 7.1 illustrates several feedback loops that operate within the scholarly communication ecosystem. At the bottom of the diagram, we see a simplified model of the traditional path of knowledge creation and discovery in the pre-Internet offline era. Here, quite simply, individual researchers and teams of researchers draw upon the canon of literature that is transmitted via scholarly communication channels. The transmission lines are clear: printed journals and books are distributed either directly to academics or indirectly via research libraries. Academics then use this "canon" to inspire new research, and finally the results of that new research feed back into the relevant scholarly communication channels. This is the traditional main feedback loop. Some portion of this scholarly knowledge is translated for public consumption (shown in the bottom right-hand portion of the diagram) by popular scientific publications and educational media, but the communication to the public tends to be a one-way process, disseminating information for public consumption.

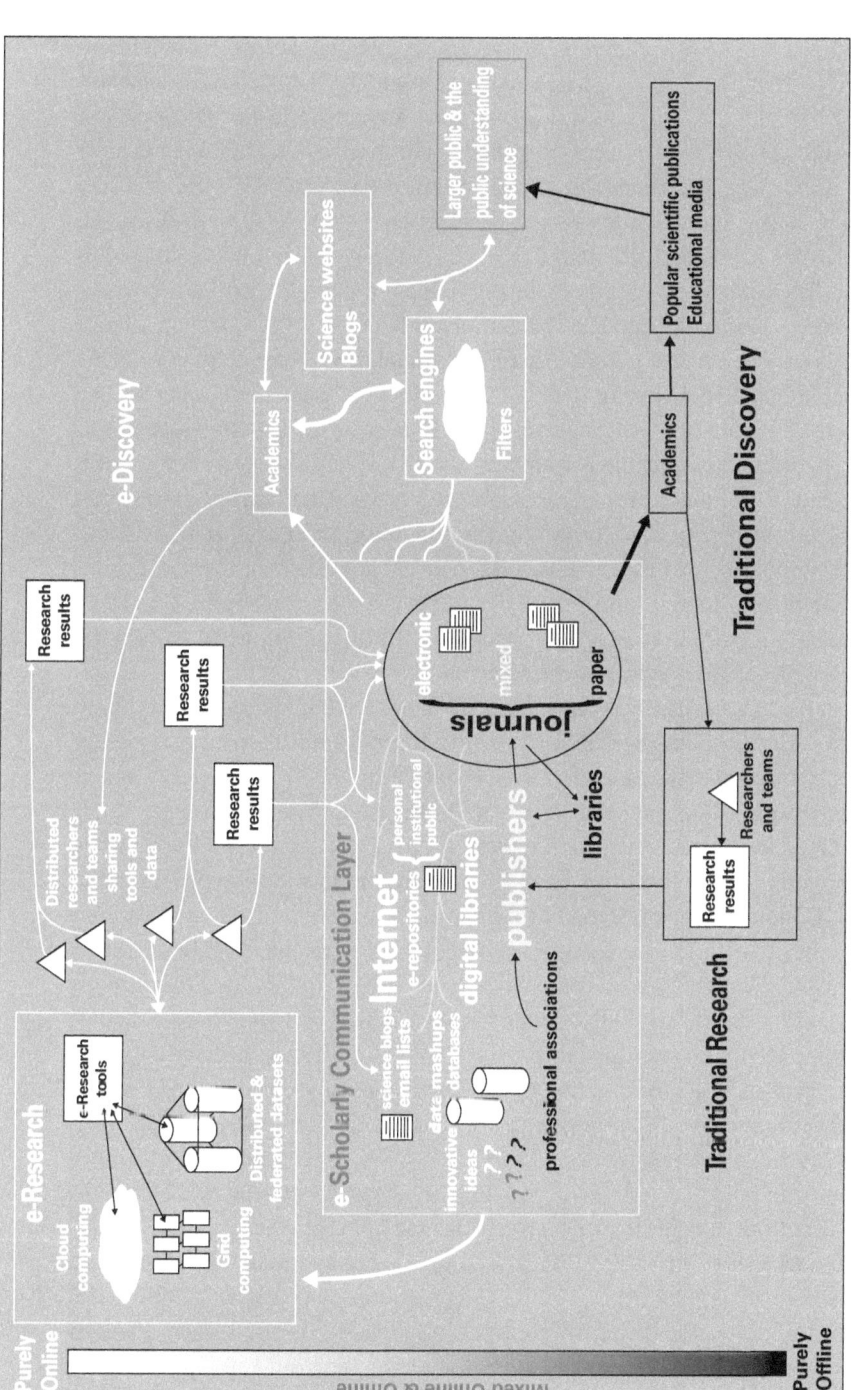

Figure 7.1

e-Research in the scholarly communication ecosystem.

Source: From Meyer and Schroeder (2009b), reprinted with permission.

At the top of the diagram is the portion of the online ecosystem in which e-research is most central. As we can see, shifts toward online scholarship result in a more complex series of interacting feedback loops. On the top left, we see the activities that encompass the research portion of the ecosystem, where shared and distributed data and tools are accessed online. These modes of access may be part of the Cloud (a term discussed in chapter 2 and derived from Cloud computing) or live on the Grid (Berman, Fox, and Hey 2003) or be part of federated data sets, and also include the tools used to manipulate and analyze data (as with visualizations of massive computations). Rather than individual researchers or disconnected teams using or generating data by themselves, in e-research collaborative efforts performed by geographically distributed teams are the order of the day. In addition, multiple researchers and teams can access the same tools and data to generate new research. As with traditional research, e-research also feeds back into the scholarly communication layer, but, as we will see, the number of ways that this can happen has increased dramatically.

In addition to the loop in the top left-hand corner in figure 7.1, there is a second, complementary feedback loop in operation, as shown in the upper right-hand portion of the diagram. Here academics are accessing the scholarly communication layer, but we also see that the paths are much more varied. No longer reliant simply on journal subscriptions and library access, academics access more scholarly literature and more types of scholarly literature. The way that they find these resources is increasingly being mediated by search engines, Google in particular. This reliance on a public search engine uncovers two unintended developments related to the Web. First, Google was not designed primarily to allow academics to find research materials (although the Google Scholar service was later designed for this purpose). Nevertheless, it has become a central tool in the researcher's toolbox for discovering knowledge (Caldas, Schroeder, Mesch, et al. 2008). Second, academic results placed on the Web were not primarily designed to be accessible to the public. Nevertheless, the fact that there is a single Internet (bar paywalls and censorship), filled with everything from humorous cat videos to peer-reviewed scientific papers on string theory, allows the public access to scholarly research in a much less mediated fashion than traditional offline knowledge. This ease of access has implications for the relationship between scholarship and the public (an issue that we turn to later in this chapter).

It should be noted that the difference between the traditional research at the bottom of the diagram and the e-research at the top is not strictly binary; there are shades of involvement in e-research along the vertical axis. However, there are fundamental differences at the two poles of the vertical axis, and evidence supports our contention that scholars in general are migrating upward in this ecosystem, away from offline modes toward online modes throughout the scholarly process. As we shall see, this shift has implications for how the knowledge within the ecosystem is managed, including in the loci of organizational control.

Decades of hyperbole about the elusive promise of the paperless office notwithstanding (Sellen and Harper 2001), the production, transmission, distribution, search for, and consumption of academic knowledge is increasingly occurring paperlessly, and this trend has been enabled by Internet-based scholarly materials. Although there remain differences in the extent to which scholars in different disciplines rely on electronic resources instead of on paper sources (Kling and McKim 2000; Fry and Talja 2004; Tenopir, King, and Bush 2004; Tenopir, King, Boyce, et al. 2005), the overall impact of electronic resources on scholars' activities is moving research in a clear direction—online—and a number of studies (in addition to our analysis in this book) document this shift.

Julie Hallmark (2004), for example, reports on a study done at two time points (1998 and 2002) that examined how geologists and chemists were finding and retrieving research articles. The method in this study involved asking scientists how they had found and retrieved a specific article that they personally cited in a publication. In 1998, whereas 83 percent of the sample of chemists used Internet-based indices to find the article they cited, only 5 percent actually retrieved the article electronically. Most still relied on paper copies of articles retrieved from libraries, personal journal collections, or the authors themselves. By 2002, however, the landscape had shifted entirely toward online search and retrieval. Eighty-five percent of chemists still used Internet sources to search for the article in question, but now 96 percent used the Internet to retrieve the article either to read it electronically or to print it out for reading. The geologists in the study followed a similar pattern. It is remarkable that, for this sample at least, nearly all articles were being retrieved from electronic sources as early as 2002.

Generational differences also appear to be at play in the changing practice of retrieving articles electronically. Nila Sathe, Jenifer Grady, and Nunzia Giuse (2002) conducted a study comparing the use of print and electronic journals. Using a small sample of journals in one library that were available in both print and electronic editions, they surveyed users of each and found that younger scholars (fellows, residents, and students) were much more likely to have retrieved the journal electronically (57–70 percent across these categories), whereas the clinical/research faculty most often used the print sources at that time (with only 25 percent retrieving articles electronically).

There are potential pitfalls, however, for younger scholars who rely on electronic resources. The highly publicized "Google Generation" project and report puts it this way:

> Most visitors to scholarly sites view only a few pages, many of which do not even contain real content, and in any case [they] do not stop long enough to do any real reading. This is either a symptom of a really worrying malaise—failure at the library terminal—or maybe a sign that a whole new form of online reading behaviour is beginning to emerge, one based on skimming titles, contents pages and abstracts: we call this "power browsing." We urgently need to understand the root causes of this phenomenon. (Centre for Information Behavior and the Evaluation of Research 2008, 31)

The report also says that "the ubiquitous use of highly branded search engines" entails, among other things, that "many young people do not find library-sponsored resources intuitive and therefore prefer to use Google or Yahoo instead" (12). "Tools like Google Scholar," they say, "will be increasingly a real and present threat to the library as an institution" (13). "Students usually prefer the global searching of Google to more sophisticated but more time-consuming searching provided by the library" (31). This report in many ways puts the lie to the meme that young so-called digital natives are sophisticated users and producers of information, seamlessly moving between online roles and connecting various electronic devices and resources through mashups and social networking. Instead, although the report acknowledges that young proto-academics may be quite comfortable with technology and tend to prefer electronic resources, it expresses serious concerns with children's and college students' ability to adequately search for and evaluate information, and it rejects the notion that the Google Generation is made up of "expert searchers" (20).

Another recent study (by the same research group) notes that "undergraduate students tend to search the Internet first, then go to library-based services, unless they have been provided with and instructed on how to use a specific resource" (Nicholas, Huntington, Jamali, et al. 2006, 1348). If this is the pathway for finding resources, however, then it is likely that different resources will be found and used than those that result from going to a library first, whether online or offline. Like the "Google Generation" report's concern with the younger generations' ability to search and find relevant information, David Nicholas and his colleagues report that when trying to use digital libraries, "web users do not dwell, they examine just a few items/pages before they leave" (2006, 1363).

Much of what is known about access to online knowledge remains anecdotal: To what extent do students and scholars rely on electronic versions of papers rather than paper versions? As Christine Borgman notes, whereas scholars may be able to make "fine distinctions ... in assessing the quality of a document ... students, practitioners, scholars with minimal access to the published literature, and the general public usually are happy to read and cite any free version of a document they can find online" (2007, 84). However, even if college students do more than simply use Google searches for sources and make use of other online sources such as *Wikipedia* as well as offline sources, the very fact that they are uncertain about which sources to use (Head 2007) suggests that visibility and access are important determinants of what they will find.

These concerns about the skills and judgment of younger generations of scholars are certainly a potential challenge for educators. They may not apply to academics who have successfully negotiated doctoral study and who will experience no particular difficulty in searching for information, even if they face an increasingly varied set of online sources. However, the preference for digital resources, especially among younger generations of scholars, is being manifested in other ways as well. Channels of communication honed through use during student years are also appearing in modified forms in the academic arena. These communications, many of which fall under the now common title "Web 2.0," include various novel forms of electronic informal scientific communication, such as blogs, personal web pages, podcasts, YouTube videos, and wikis (Nentwich and König 2012). These forms are now being added to existing informal modes of academic communication, which include e-mail, e-mail lists, conferences, and professional newsletters. Although some forms of electronic communication are technological replacements of previously existing modes of communication (such as email replacing snail mail), the same is not true of many of the latest innovations. Blogging does not have a clear analogue in the paper-based world; journal writers may have kept track of their thoughts on a variety of topics, but they did not post them publicly unless they published them as memoirs at a much later date, often after their retirement or death.

The shift of research materials online thus involves a variety of informal means of scientific communication, and although it will not be possible to review all of them, a few examples can suffice. Mike Thelwall and Kayvan Kousha (2008), for example, examined PowerPoint presentations available on the Web looking for evidence of whether they could be used as nontraditional indicators of research impact. Using a combination of automated searching for PowerPoint files containing references to Institute for Scientific Information (ISI) journals and manual classification of an additional sample of presentations to look for other types of citations, they found that, in general, not enough presentations contained sufficient journal references to make the construction of an impact measure worthwhile. They did note, however, that online presentations often cite more popular resources such as *Scientific American* and *Harvard Business Review*, which could be useful for tracking the popularization of research. Likewise, David Wilkinson and his colleagues (2003) also found that informal types of scholarly materials dominate the websphere. They extracted web-link information from 107 university websites and found that almost 90 percent of the links were created for scholarly reasons, but that only a minority consisted of links to journal articles that could be considered the equivalent of a citation. The other links led to a variety of materials, including information for students (18 percent) but also material related to research resources (17 percent) and to libraries and e-journals (15 percent).

In general, if scientists can be understood as classic maximizers, they should engage only in alternative methods of informal communication if those methods are better in some measurable way to existing modes of informal communication. Uwe Matzat's (2004) study of scientific Usenet groups as channels for informal scientific communication, for example, concluded that although there was little support for the hypothesis that such groups had a democratizing effect on scientific communication, he did find evidence that participants in the groups reaped benefits in terms of research information and in maintaining weak ties with members of their extended networks.

Understanding that scientists communicate scientific information informally is not a new observation, but it is worth noting that the channels with which they do so are proliferating. William Garvey and Belver Griffith pointed out several decades ago that "scientists themselves create elements to fulfill the information needs that are not being satisfied by existing media" (1967, 1012) and argued that these new elements would evolve over time and result in a shift in norms within a scientific field. Their example of certain fast-moving disciplines adopting increasingly speedy methods of exchanging preprints shows that tools such as arXiv.org are far from being completely new, technologically mediated innovations but instead are the current incarnation of a trend noted in the literature 40 years ago (1967; 1972). Specific scholarly behaviors may have changed, but the overall "socioecological system" (Sandstrom 2001) of scholarly communication has continued to evolve along long-established lines (Hakken 2003; Heimeriks and Vasileiadou 2008).

For formal and informal academic materials to have impact, they must be visible to their potential audiences. This is one area where the Internet offers much greater potential than the library-based paper publishing system ever did. Once academic material is on the Web, particularly if located in open-access sources indexed by Google and other search engines, other scholars and members of the general public at least have a chance of finding the material. The apocryphal story of the doctoral dissertation on the library shelf still containing the $20 bill placed there decades earlier by its author reflects the understanding that few would bother to access something difficult to find and of such limited interest. Putting the same dissertation online doesn't make it more interesting to a wider audience, but it does make it much more likely that if someone *is* interested, they may look at it on the Web. The same is true of a variety of other academic outputs.

With regard to access by the general public, for much scholarly work there will always be a quite limited public audience. Nevertheless, the Internet is not compartmentalized and divided into separate physical spaces the same way that public libraries and academic libraries have been in the past. By mixing one's academic work in with the other material in the cloud of information that everyone uses on the Internet, it becomes more likely that others may stumble on that work than if it is locked away in dusty, little-visited academic libraries. Borgman puts the issue succinctly: "Content that is online gets more use than that which is not" (2007, 159). And, as Gaston Heimeriks and Eleftheria Vasileiadou point out, "A scientist's visibility does not rely exclusively on the number of publications and their citations but can increasingly result from a well-designed and well-linked homepage providing scientific content" (2008, 18).

Placing academic content online and allowing it to be freely used by others is referred to as "open-access publishing" (we discuss this openness in the next chapter). Borgman argues that enhanced visibility is one of the main motivations for open access (2007, 101). However, how individual researchers' visibility is related to their impact as scholars is still unclear. Franz Barjak, Xuemei Li, and Mike Thelwall's (2007) study of inlinks to scientists' personal web pages found conflicting results. They found that full-text articles were the most linked content on the personal pages, but they found disparate results when examining the scientists' collaboration networks. Contrary to expectations, having a large number of collaborators had a negative impact on the number of inlinks, and productivity was similarly not reflected by the number of inlinks. The authors conclude that the main lesson is that our understanding of the role of visibility on the Web is still incomplete (for more on this topic, see also Houghton and Sheehan 2006).

If one accepts the premise that online visibility is a growing trend, it becomes possible to speak of the online "presence" or "visibility" of research. However, presence and visibility have not yet been well defined; a conception of the Web as a whole, of the Internet and the Web as a *system*, is currently missing from the debate, and thus there are insufficient models for understanding competition for attention within this space. The notion of the "websphere" (Schneider and Foot 2005) has been used, for example, for analyzing political phenomena, but it is unclear if this notion can be applied to research outputs, which are more varied. The reason for the lack of an understanding of the competition for visibility is also that much of the focus has been on the producer side, on outputs or on those who put information and knowledge online. Equally important, however, is the consumer side: How are the use patterns in researchers' information and communication practices changing in light of the other changes we have discussed in the scholarly communication system?

Moreover, in relation to both production and consumption of online knowledge, there has been a focus on the question of whether the shift of materials online results in a winner-take-all system (the "Matthew effect") or in a more democratic system of scholarly attention (Caldas, Schroeder, Mesch, et al. 2008). As we shall see, this question may be too limited. Yet in either case the competition to dominate the attention space has moved onto the new terrain of the websphere, and, as a result, new players have entered the equation and have altered the role that different mechanisms—such as search engines—play in determining scholarly visibility.

e-Research contributes to shifting knowledge online and by definition produces materials for online access. However, the visibility and dominance of online resources must also be seen in a context that is larger than searches, fields, and formal and informal scholarly communication. Rob Kling, Geoffrey McKim, and Adam King have suggested that it is possible to see new electronic forms of scholarly communication such as "electronic editions of paper journals, pure electronic journals, working article repositories, post-publication archives, pre-print servers, collaboratories, cross-linked Webs of resources, gene databases," and the like as part and parcel of a set of e-scholarly communication forums (2003, 47). They go on to point out that this does not mean that these forums are therefore purely electronic because researchers will also continue to exchange information face to face.

Nevertheless, these e-scholarly communication forums could be regarded as overlapping with emerging e-infrastructures or cyberinfrastructures, but the latter also constitute something larger. e-Infrastructures are systems of networked digital resources that will serve fields but also scholars across fields at a national or supranational level in the manner of a long-term support mechanism to support research (Borgman 2007; Schroeder 2007a). Thus, the shift to online resources cannot be left on the level of scholarly communication practices but must be raised to the level of transformation in the very systems of scholarly communication. Jenny Fry (2006) uses the term *scholarly networked digital resources* to refer to the overall system beyond individual projects, digital libraries, and discrete web pages. A broader conception such as this allows us to include in the discussion both the infrastructure and its networked parts that make up the scholarly online ecosystem.

The online ecosystem thus consists of more than just scholarly communication. Within scholarly communication, a distinction is made between, on the one hand, formal communication, which is long lasting and addressed to a wider audience, and, on the other, informal communication, which is more ephemeral, and between public communication and private communication. However, as Borgman points out, these lines are especially hard to draw with digital scholarship (2007, 48–49). Hence, Borgman also notes that "in digital environments, dissemination can be difficult to distinguish from access" (2007, 87). In other words, what is found online can be regarded as published for dissemination, and what is disseminated can be seen as being published, even if this interchangeability was not intended.

However, Kling and McKim (2000) are skeptical of overly optimistic views of the power of information and communication technologies to transform scholarly activity. Their central argument is that although there are many examples of increasing digital scholarship, important field differences remain and shape the extent to which disciplinary actors are likely to actively engage in e-research. (We use the term *field* here when describing an emergent scholarly domain, such as a specialism or multidisciplinary effort, that does not fit within the boundaries of existing disciplines.) For instance, in computer science and physics, articles are published online; and as we have seen, computer scientists have been particularly aggressive in terms of placing their papers on their personal websites (Borgman 2007, 102). Does this mean that these two disciplines will become more visible on the Web vis-à-vis other less-aggressive fields? This would follow if we combine this trend with what Stevan Harnad (2001) has noticed about how much more open-access material is cited—but only if knowledge transcends disciplinary boundaries; if it does not, then the visibility should be "contained" within the two disciplines.

Disciplinary differences can be overemphasized. Jonathon Cummings and Sara Kiesler (2005) found, for example, that cross-disciplinary collaboration is not so much of a problem for collaborative research as is research that spans across different institutions or that bridges distances (in other words, with distributed teams). Similarly, John Walsh and Nancy Maloney found that the structure of work, including size, distance, interdependence, and scientific competition, are more problematic for collaboration in scientific teams than the mix of backgrounds that is often assumed to be problematic (2007).

Nevertheless, disciplinary differences are evident when it comes to the speed at which the processes described here are occurring. Fry, drawing on Richard Whitley (2000), has argued that, in terms of "the differential role of informal and formal communication across fields," the characteristics that Whitley ascribes to different fields have "an influence on the production and use of scholarly networked digital resources" (2006, 312), so that fields with a high degree of mutual dependence and low degree of task uncertainty, such as high-energy physics, are much more likely to produce and use these resources than fields with a low degree of mutual dependence and a high degree of task uncertainty, such as social/cultural geography. There are, however, patterns that override such differences; for example, researchers in all four fields examined in one study (terrorism, HIV/Aids, climate change, and Internet research) use scholarly networked digital resources in such a way—searching with Google—that they are ever more reliant upon these resources (Fry, Virkar, and Schroeder 2008). If you search for an individual scientist's personal web page, visibility will be even more important in a field that makes low use of these resources. Other scholars have also looked at discipline-specific use of scholarly networked digital resources, comparing, for example, astronomers (Tenopir, King, Boyce, et al. 2005) and medical faculty (Tenopir, King, and Bush 2004). It is clear, therefore, that field differences will persist in terms of the extent to which electronic scientific communication is adopted (Kling and McKim 2000; see also Walsh, Kucker, Maloney, et al. 2000). Yet even if humanities and social sciences "lag," this lag must nevertheless be put in the context that all disciplines are moving in the direction of digitizing online resources. The key point is that *all* disciplines are doing this in different ways and will thus be subject to the competition for visibility that we outlined earlier, even if this competition will take various forms.

The very idea that commercial search engines, a dominant one in particular, should be used to access scholarly knowledge would have been unthinkable 10 years ago. Google and Google Scholar are increasingly playing a gatekeeping function in e-research (Fry, Virkar, and Schroeder 2008). Google has actively moved into an area formerly dominated by players such as Thomson/ISI. It would be hard to argue that the former gatekeepers were terribly democratic in their policies toward access to knowledge: most of these proprietary databases were carefully locked behind subscription walls raised and guarded by university libraries. Google Scholar, however, is not a database but an index and search engine. The articles that it finds are still often locked away behind subscription-based interfaces unless one is accessing them from a research university that can afford to subscribe to these resources.

Further, we can ask: How well does Google Scholar fare at finding relevant, high-quality scholarship? Lokman Meho and Kiduk Yang conducted a study comparing Google Scholar's ability to find citations to an author's work to Scopus's and the Web of Science's ability to find the work of a single highly cited library and information science department. They determined that Google Scholar was superior to the other sources at finding citations to an author's work in conference proceedings (four times more), non-English sources (more than six times more), and works self-archived on a personal or institutional website (which are not covered by the other sources at all). Thus, authors who choose to self-archive have a "dramatic advantage" in terms of their visibility in Google Scholar (2007, 2118). In terms of accessing peer-reviewed journal literature, however, Scopus and Web of Science are better at filtering out low-impact sources of citations; most of the additional citations Google Scholar was able to find were from low-impact journals and conference proceedings, and their inclusion did not change the relative ranking of scholars' productivity. The wide coverage that Meho and Yang report for Google Scholar is in marked contrast to the findings of an earlier study, that Google Scholar suffered from "massive content omissions" (Jasco 2005, 208).

Kayvan Kousha and Mike Thelwall (2007) compare ISI citations to Google and Google Scholar using a different methodology. They similarly found variability in the effectiveness of Google Scholar, noting that fields with a bias toward valuing conference articles and placing them online, such as computer science and some social science disciplines, were better represented in Google Scholar. Overall, however, they found a strong correlation between Google Scholar citations and ISI citations.

Meho and Yang point out that some structural problems with Google Scholar limit its use as a bibliometric tool but also have implications for scholars who rely on it as a tool for research: Google Scholar lacks full bibliographic information and metadata on the sources it finds. More importantly, Google Scholar "ranks the items in a rather inconsistent way … [and] does not allow resorting of the retrieved sets in any way (such as by date, author name, or data source" (2007, 2110–2111). However, if one wants to find more marginal literature (such as that located in what Chris Anderson [2006] has called the "long tail"), Google Scholar is a better source than Scopus or Web of Science. There are many reasons why a scholar might be interested in moving outside the mainstream journal articles: among others, to examine an underresearched topic, to find references to a newly emerging topic that has not yet had time to appear in mainstream publications, or to find international perspectives on a topic of interest beyond the Global North.

Hence, if we return to figure 7.1, we can now consider the changes in the scholarly communication ecosystem brought about by the various factors we have discussed. In the middle of the diagram, at the e–scholarly communication layer (e-SCL), we see a variety of these formal and informal modes of scholarly communication. The variety of forms that research results can take has grown enormously. Rather than having a small range of options for submitting results for publication, the online e-SCL enables multiple fluid paths for different outputs and even for copies of the same output. Researchers can and do still publish their results via the traditional routes (articles in peer-reviewed journals and books), but they can now also take those same results and post about them on their blogs, send them to large groups of researchers via e-mail lists, and post preprints on their web pages and in institutional or public repositories. From the researcher's point of view, these options complement the "official" publication of results in a journal of record and increase the visibility of the results (as discussed earlier). Publishers, however, are struggling to maintain their position as these freely available copies of the material for which they charge access proliferate.

The examples toward the left in the e-SCL box are generally newer arrivals on the scholarly communication scene, whereas those on the right represent more traditional forms of scholarly communication. Objects and actors in this layer are represented in white if they are primarily online, black if they are primarily offline, and gray if they are mixed. (Paper journals are included here because, even though they fall outside the "e-" portion of the e-SCL, they still have an influential role in e-SCL as a whole.)

The Internet has played a major role in enabling researchers to engage in more widely disseminated forms of informal communication; some of these forms are enhancements of pre-Internet behaviors; others are novel forms of informal communication. Furthermore, the online elements have fewer direct lines connecting them. Instead of direct relationships between, for instance, publishers and libraries negotiating subscription terms, the online elements are tied together in a web of relationships that connects them to each other. These relationships can obviously be as simple as hyperlinks, but they can also be as complex as the institutional demands of organizations such as the US National Institutes of Health to make funded research results available via freely accessible channels.

On the right-hand portion of the diagram is the process of discovery of research results. This process is made up of the search engines (discussed earlier), which are one type of filter, along with elements such as science websites and blogs geared to the general public, which also offer access to scholarly material. Researchers may access materials directly, as indicated by the arrow directly from the e-SCL to academics, but they are generally likely to access materials through one or more filtering mechanisms. Of course, academics may also be those who produce scholarship, as indicated by the arrow back to the distributed researchers, but they could equally be those accessing knowledge that is outside the domain where they are the producers of knowledge. The barriers between fields and disciplines, although still strong in practice, are lowered for those who wish to engage with knowledge and data from outside their field. This is a serious challenge for those interested in the dissemination of knowledge within and between organizations (such as publishers or those interested in scholarly communication for particular fields or disciplines): the fluid paths to knowledge are difficult to manage, particularly because the Web itself is constantly evolving and shifting as resources appear, move, and disappear over time.

Note that in the diagram the organizational effects of being located to a greater extent in the online space (toward the top of the diagram) are not straightforward. As discussed earlier, the e-research efforts in the top left can either be much more organizationally structured than the traditional research at the bottom (as in the case of large e-infrastructures) or be much less structured, as in the case of Web 2.0 approaches to research. Similarly, in the right-hand portion of the diagram, the examples of traditional paths of discovery and communication with the public at the bottom are highly structured and dependent on organizations, whereas the paths to accessing information and knowledge at the top portion of the diagram are much more fluid, available through multiple paths, and generally more openly available. However, to ignore the fact that Google itself is a large organization and has corporate policies that influence what information is available to searchers would be naive. Indeed, although Google and other search engines in some ways make online information more transparently available, the actual workings of how they do this and how results are ordered are not at all transparent to the average academic or citizen. The search engine has been black-boxed, and the results that pop out of the box are taken as a given.

The wider public obtain their understanding of research both through researchers who engage in expanding the public understanding of science and increasingly through direct access to information through filters (such as Google), which are widely available to those without university access to resources. This filtered—and in other ways less filtered—access is a new phenomenon; other than the occasional enthusiast willing to go to an academic library and make photocopies of research articles, the general public in the pre-Internet age had very little access to research material outside of popular magazines and television programs. Such access has also introduced a tension, however, particularly in some sensitive fields such as medical research, where untrained readers may misinterpret scientific information and place their health at risk. Doctors have complained in anecdotal reports of increasing numbers of patients coming in to consultations armed with printouts of information from the Internet that the clinician considers to be of dubious quality or inaccurate. In this respect, the role of scholars as the sole gatekeepers to the interpretation of scientific and other information has been weakened.

Conclusion

In chapters 6 and 7, we have covered a great deal of ground to help demonstrate how disciplinary differences are enacted with relation to digital research. Across the sciences, social sciences, and humanities, we have seen examples of scientists and scholars who have engaged in e-research either as an active choice to advance science (EGEE, IVOA, VOSON, culturomics/distant reading, SwissBioGrid, Swedish data sharing) or by accident as they pursue research questions and find that digital tools and materials are an effective path toward answering the questions they are interested in (Google search data, digitized humanities collections, Galaxy Zoo, *Pynchon Wiki*, SPLASH, and GAIN).

In all these cases, computational or digital approaches to research are gaining prominence across academic disciplines, albeit unevenly and in the face of the persistence of disciplinary norms and values that are transmitted and learned at an early point in new researchers' careers. However, the entire ecosystem of research, scholarly communication, and dissemination and discovery is changing, and disciplinary responses to these changes are still evolving. We can finally note a further all-encompassing, if also still evolving, shift: the knowledge that has moved online can be gauged more powerfully because it leaves digital traces. Put differently, because digital objects that are the outputs and traces of research are online, they can be analyzed and measured using digital research technologies. For example, web links are analyzed using tools such as VOSON; traces of PowerPoint presentations are measured by scraping websites; journal reading habits are analyzed by using logs; search engine and other logs tell us what researchers and others are reading and what they are interested in (Van de Sompel, Payette, Erickson, et al. 2004); relations between disciplines are charted by means of citation analysis; influence is measured by analyzing links between researchers' blogs; and so on. Thus, researchers are much more aware of the impact and visibility and shape of their research and increasingly orient themselves toward this impact and visibility in a feedback loop. Put even more abstractly and in keeping with our argument about advancing knowledge and research technologies, the very machinery of research and its shifting nature are becoming more visible and measurable, with implications for the role of knowledge in society, a point to which we return in the conclusion.

8   Open Science

It has been argued that open science is a precondition for the whole of modern science (David 2004), but this argument has taken new forms in the case of e-research (Schroeder 2007a; David, den Besten, and Schroeder 2010). The push toward openness and against it is currently coming from several directions: in publishing, funding bodies are trying to mandate open (often gratis) publication, including data, to maximize return on the taxpayer's investment. At the same time, many publishers are fighting against this trend in order to maintain profits from professional journals. Among nongovernmental organizations (NGOs), there are various initiatives to promote a more open intellectual property regime that would, for example, benefit the fledgling biotechnology industries, especially in developing societies (Hope 2008). On the other side, industry lobbies have been pushing for more restrictive intellectual property rights as part of a wider effort to lock down the sharing of digital materials and to monetize the fruits of basic academic research. And finally, open-source software constitutes another push because the software for e-research is often licensed with an open-source license, and efforts are being made to maximize the user community with interoperable standards (Weber 2004). However, these efforts suffer from confusion concerning which licenses are best suited in different circumstances (Fry, Schroeder, and den Besten 2009) as well as from fights about standards among competing groups that promote different types of software within a shifting ecology of interconnecting parts and rival models (such as web services versus dedicated Grid software, recently joined by Cloud computing).

This multitiered battle between more open and closed forms of e-research is part of a larger struggle between different ways of pursuing knowledge, and this struggle is not and will not be decided with one model triumphing over the other. Instead, there are bound to be variations within and between fields and domains (for example, government and the private sector or biotechnology and the software industry), depending, if this applies, on factors such as the commercial value and copyright legacy of the knowledge concerned. Even here, however, certain new rules of the game (which also apply in other areas, such as business and consumer software development) affect e-research in particular—for example, the notion that locking in a large user community with free access may be more valuable than charging for subscriptions to a database. These new rules are still fluid, so that the sustainability of open-science models remains in question, though there is increasing pressure for open access to born-digital research materials (Borgman 2007).

Pushes for Openness

Various institutional efforts have been made to promote openness in e-research.[1] For example, the Open Middleware Infrastructure Institute[2] was an effort (ended in 2010) to provide and support free, open-source software for UK researchers; likewise, Jisc (formerly Joint Information Systems Committee, or JISC), an umbrella body that provides information and communication technologies for the United Kingdom's education and research sector, originally funded the "OSS Watch,"[3] the "Open Source Advisory Service," which promotes open-source software in UK higher education, although Jisc's funding support ended in 2013. In addition to research funding bodies, NGOs have also pushed for open science on the level of research policy. One example will suffice: Science Commons[4] is an NGO whose aim is advocacy on behalf of open science. Science Commons comes under the umbrella of the Creative Commons organization, which promotes copyright licensing that is open in the sense that it offers commons[5] alternatives for those who might otherwise use conventional copyright and instead uses licensing that preserves open access. In other words, it promotes a form of licensing that is an alternative to for-profit copyright and patenting, thus making it easier to share scientific knowledge without forcing it into more exclusive copyrights or patents. In fact, there are now several different types of creative commons licenses, which differ in the strictness with which this "openness" or "freedom" is interpreted. Science Commons is one among several creative commons initiatives, and it is subdivided into several more specific areas of concern (publishing, data, and licensing). What unites these specific areas is the aim of promoting open access to and use of scientific knowledge.

A number of institutions such as publishing repositories and NGOs support and have adopted Science Commons aims. A related effort, for example, is the Global Information Commons for Science Initiative by the International Council for Science's Committee on Data for Science and Technology, which aims to promote better understanding and coordination toward an "information commons" for global e-science (David and Uhlir 2005). More recently, there has been a push for open data with the Research Data Alliance.[6] Many other examples of openness policies among research funding agencies and NGOs can be given, both within the United States and the European Union and beyond. It can be seen that these policies are almost identical in their thrust and aim to promote openness in all the senses mentioned earlier.

Science and the Norm of Openness

Although *open science* is a relatively recent term, it is possible to argue that modern science has always been open (David 2004). Robert Merton proposed in the early 1970s that the "ethos of modern science" consists of "four sets of institutional imperatives—universalism, communism, disinterestedness, organized skepticism" (1973, 270). In the context of open science and e-research, the most relevant of these imperatives is communism, by which Merton did not mean the political ideology, but rather more generally the "common ownership of goods." He argued that "the scientist's claim to 'his' intellectual 'property' is limited to that of recognition and esteem which ... is roughly commensurate with the significance of the increments brought to the common fund of knowledge" (1973, 273). Within the sociology of science and technology, Merton's account of the norms of science has been extensively criticized. David Hess has reviewed these criticisms and concludes that "it is possible to salvage Merton's delineation of the norms of science, but only as a prescription of how scientists should behave ideally" (1997, 57); in other words, these norms do not reflect actual scientific practice. Others have argued, however, that, with some refinement, they generally *do* reflect it (David, den Besten, and Schroeder 2010).

In the case of e-research, these norms of scientific inquiry are perhaps less important than the fact that they underpin a system of knowledge generation and sharing in a broad sense that includes not just publication but also access to the tools for research. In this respect, Merton's norms of "universalism" and "communism" are part of the broader openness of the research process.

Apart from Merton, other sociologists of science have argued that openness is an essential institutional characteristic of scholarly communication and thus a necessary condition for the advance of scientific knowledge. That is, scientific knowledge must remain open to being improved upon via new communication because closing off communication would prevent the ongoing refinement of knowledge (Becher and Trowler 2001; Fuchs 2001b). This fundamental sense of openness relates to the conception of science and knowledge presented in chapter 2 and applied in subsequent chapters: science as an ongoing and open-ended process of advancing knowledge and more powerful mastery (manipulation) of the physical world—a process that, as Stephan Fuchs (2001b) argues, is intrinsically open in the sense that scientific knowledge must not close itself off (for example, by insisting on a particular normative stance) to testing and evidence.

The question whether e-research and e-infrastructures are open—in the sense of the "communism" of shared resources and the "universalism" of open communication—and live up to the institutional norms of science is thus a key question.[7] Apart from research policy and the norms of science, there is also, however, a much broader impetus toward openness. In fact, so many initiatives for openness exist now that it is difficult to keep track. Thus, openness movements can be found in the realms of culture and media (for example, open publishing), politics and political activism (for example, open data and transparency), and business and economics (open innovation).

The most well-known example of such an initiative is open-source software. Open source is the attempt to create software that can be freely used and developed within terms of a license that safeguards continuing free use and access. A fundamental distinction is between software that is purely noncommercial (i.e., it contains no proprietary code and stipulates that all code must remain nonproprietary) and software that allows proprietary— or commercial—appropriation and development. The advantages of different models—whether it is best to develop so that whatever is developed can be commercially exploited or to stay "pure"—has been subject to extensive debate.[8]

Open source is critical to e-research because open-source software tools are important shared resources in many of these efforts. Open source, according to Steven Weber (2004), has gone furthest in addressing some of the legal and business-related issues that are also relevant to e-research. One question is therefore to what extent open-source collaboration and software practices can provide a model and viable tools for open e-infrastructures and for the openness of e-research.[9]

Other initiatives promoting open access to knowledge, especially in publishing and related areas, are too numerous to chart.[10] They pertain to libraries, texts, data, and other digital objects. John Willinsky (2005a) also points to the similarities between open science, open access, and open-source software. Much like open-source software, he argues, which is open to contributions and continual refinement from a larger community, research, too, can be seen as a product of continually building on other researchers' openly published achievements and improving on them. Willinsky concludes that "the current convergence among various *open* approaches to intellectual property represents a common commitment to a larger public sphere. These approaches extend well beyond the university and yet it remains the primary institutional force in sustaining this open economy" (10). At the same time, the idea of collecting these efforts under one umbrella is not intended to give the impression here that they are part of a concerted or centrally organized effort. In fact, a key feature of open science is that it consists of disparate initiatives, mainly by researchers, academic institutions, and NGOs.

Against the spread of open-research policies and related initiatives, there is also a contrary trend toward securing intellectual property rights in such a way that ownership is protected—in other words, that access to data and publications, the use of software and tools, and also standards should conform to a proprietary or exclusive regime.[11] This trend is sometimes referred to as the new "enclosure movement" (Benkler 2006) on the analogy with what is often regarded as the origin of private property in the first enclosure of public (commons) agricultural land.

The issue of either protecting or unlocking intellectual property rights has moved to center stage for a number of reasons. One is that whereas patents and copyright were previously subject to national jurisdictions, with international agreements in place for enforcement beyond borders, it has in recent years become necessary to ensure that this enforcement extends globally and especially to the developing world and for those forms of intellectual property such as patents for medicines and copyright for entertainment (music and video material), which may not have enjoyed this protection or where it is being challenged. Another reason includes increasing pressures to commercialize research (particularly in certain domains such as biomedical research or in certain areas of engineering research), including research-funding pressures for academic research to produce "added value."

Some have argued that a narrow constituency of private interests has been winning the battle to enshrine global enforcement of commercial intellectual property rights (Sell 2003). At the same time, as we have seen, moves toward open research and openness in other domains have also been gaining strength. To be sure, the two do not necessarily conflict or confront each other directly, so that proprietary commercial research, for example, can coexist with openly accessible research results on the same topic. At the same time, a battle is currently shaping up between "two distinctive regimes or environments for the conduct of research: the actors in the realm of 'open science research' expect reciprocal sharing of discoveries among themselves and the rest of the world, while those in the world of private profit-oriented and proprietary R&D expect to receive payment for the right to use their inventions (and to pay others for the use of theirs)" (David and Hall 2006, 767; see also Burk 2007). A key question is therefore whether those building tools and infrastructure in support of e-research will design for openness.

The movement toward openness in research can thus be seen as a legal battle, but it also resonates with a larger ideological or cultural movement that is shaping the agenda about globalization, social justice, and common resources. This larger environment is bound to affect policy and practice in the research community and the technological infrastructure that supports it. One reason why a wider social context is crucial is that although science has been an autonomous institution in society since the late nineteenth century with the emergence of the modern research university, in the economic and political climates of the late twentieth and early twenty-first centuries science needs allies (Collins 1993). It needs allies in part because the equipment for carrying out scientific research in the age of big science and large technological systems has become very expensive and in part because of growing controversies in certain areas of science.

A number of actors populate this larger social movement toward openness. Most narrowly, they consist of researchers, and Peter Haas has coined the term *epistemic communities* to characterize the social role of experts and expert activism. He defines an epistemic community as a "network of professionals with recognized expertise and competence in a particular domain and an authoritative claim to policy-relevant knowledge within that domain or issue-area" (1992, 3). More broadly, these experts can also be seen as "activists beyond borders" who are networked within NGOs to promote social justice (Keck and Sikkink 1998). And more broadly still, they also include the midlevel government officials whom Anne-Marie Slaughter (2004) describes and who, she argues, constitute a new force in international relations that is building global networks to make internationally binding agreements—thus exercising "soft power."

The movement toward openness aims to maximize the benefit of research, and in this advocacy mode researchers and others will stress the social and economic benefits of science and research for society. In doing this, they are supported by wider movements in society and by a public receptive to science that is seen as a public good. Whether the engagement between science, the public, and the wider society in this case should be regarded in terms of a conflict or controversy (mainly over resources), as a conflict sociologist might see it (Collins 1993), or in terms of a functionalist framework whereby interest groups and the research community need to be aligned in a non-zero-sum game (Bauer and Gaskell 2002) remains to be seen.[12] In assessing the significance of the agenda of this global social movement, we must recognize, however, that even when NGOs and the scientific community and social movements operate on a worldwide basis, the target of these movements is nevertheless primarily the state.[13] This argument applies to research in particular because the state is the main source of funding of academic research and the source of research regulation.

The Limited Impact of Open-Research Infrastructures

There are several limitations to the extension and impact of openness within e-research. The most important of them is the limitation on the "attention space." The concept of attention space on the research frontier (which we have already discussed in passing) has been used to analyze intellectual change as a whole (Collins 1994, 1998), but it is necessary to adapt this notion to the realm of e-research. This can be done by noticing that whereas attention space applies to knowledge, similar ideas have been applied to technology under the labels *lock-ins*, *path dependence*, and *monopoly*.[14] And although these ideas are not identical, they all go against the thrust of a characteristic of openness that is often taken for granted in the debates discussed so far: namely, that it is possible to create new systems without displacing existing ones and that open research is infinitely expandable and yet imposes no restriction on other systems or parts of systems.

In the world of research, moreover, the status order that governs what counts as knowledge relates not just to the *amount* of attention, but also to *who* pays attention. In academic research in particular, this status order is well entrenched, with certain researchers or research groups attracting the bulk of attention—also known as the Matthew effect[15] (see Merton 1968), which has a tendency to lead to a "winner takes all" scenario in which initial prominence disproportionately influences later prominence. With respect to e-research, the same applies to electronic networks: these networks must lock users (or "customers") in or dominate the "attention space" to obtain competitive advantage in the market for symbolic goods or tools or materials—in this case, scientific knowledge or instruments or data sources in their various forms. Debates about the "sustainability" of e-research can be interpreted in the light of this limitation: only those tools or projects that gain enough attention and users or those tools that are deemed useful for making research advances can be sustained with resources.

This aspect of e-research and e-infrastructures points to a difficulty: how to gauge to what extent these initiatives lack users and will atrophy. Many types of e-research will be eliminated as the result of lack of attention from users and the inability to sustain themselves with sufficient resources, as happens to much research and new technology generally. "Sustainability" has been and continues to be a topic that is often discussed at the conferences dedicated to e-research. This limitation, however, is also useful for allowing us to see the flipside more clearly: the success of openness will depend on generating and sustaining the largest possible user or attention base.[16] Again, we return to the point that certain areas of scientific advance are tightly coupled to—or highly dependent on—the research instruments they use (Whitley 2000). In areas that are more loosely coupled, it may be the case that a number of these instruments can exist in parallel, but they may also be harder to sustain if research does not depend on them.

At the same time, the attention space in academic research is not only restricted in a zero-sum way. New technology creates spaces and newly expanding domains (for example, new specialisms or subdisciplines—of which e-research or e-science is itself one) that will, again, complement and add to existing ones—but these spaces fill up. In the case of open-source software, for example, it can complement existing software and expand to provide solutions to problems not yet covered. Similarly, in academic research, it is possible to have increasing rates or volumes of knowledge production and increasing specialization of research outputs for e-research to expand and extend into (for example, new online databases). The limitation is nevertheless the focus of attention on a few research results, and this effect persists online. Specifically in relation to e-research and e-infrastructures, there will be one or a few dominant databases or sources, and, similarly, one or a few tools and types of software will dominate. Even open-access findings in e-research will be governed by the limited online attention space within different disciplines or knowledge domains.

Limited attention space is thus the main limitation on open e-research, but there are a number of further and related limits. One is not to overestimate the impact of open e-research, which an "advocacy" perspective tends to do.[17] For example, researchers mainly build on each other's work, and because most research is done in institutions where researchers have access to the resources they need—whether they are published in open-access journals or in expensive subscription-only journals makes little difference; a shift toward openness will have a limited impact on their output. Note, however, that there may be a network or bandwagon effect: if open-access research is cited more because of how researchers access such material in practice, this increased citation will reinforce the drive to publish in open-access forums. Nevertheless, the much-discussed benefits of reaching a public outside of research institutions or reaching researchers in the developing world (Willinsky 2005b) will have only limited impacts.

Another limitation (already mentioned in passing) is that this system is not entirely new. The emerging e-research infrastructure is being grafted onto existing infrastructures—of electronic publishing, databases, instruments, networks, and the like. Like other new technologies, the technologies of the e-research infrastructure complement and add to rather than replacing and superseding existing technologies.[18] Nevertheless, the e-research infrastructure is new insofar as it consists of more powerful networks and computing tools, and therefore the openness of these networks and tools or of this new part of the system will make a difference in use: researchers and others will increasingly come to take for granted that they can go online to access the materials they need (as discussed in chapter 7)—including publications, data, tools for manipulating data, instruments, and the like—but their capacity to do so will be amplified, not revolutionized, by means of the new (more or less open) e-infrastructure.

Yet another limitation is that although accessing and contributing content, tools, or other materials to the e-infrastructure may be open to all, organizational resources are needed to contribute actively to these efforts, and this requirement will impose constraints: ensuring the quality of content, software formats and standards, and accessible portals—to name only a few such affected elements—are subject to the limited resources that can be brought to bear on these tasks. And the electronic network of the Internet and the Web, which consists of different parts, needs to have the capacity to allow access, provide storage, and be maintained—again, a question of resources.

Finally, e-research is concentrated at the leading edges of scientific and research advances, but this position constitutes a limit too: the major effect of e-research will be where research advances are highly dependent on more powerful research instruments or sources of digital data and research materials—for example, in biotechnology with its impact on health, food, and perhaps related areas, such as energy. There will be impacts on other scientific and knowledge areas of advance, but we have already mentioned the well-known constraints on the intellectual and social organization of the sciences: whether scientific and other academic disciplines are closely coupled and mutually dependent and to what extent they are driven by research instruments and data. In other words, there is variation in the extent to which the advancement of scientific and knowledge relies on computing tools (or, in this case, e-research tools) and research materials (data) and how these more powerful instruments yield scientific and knowledge advances that impact on society.

Conclusion

As we have seen, openness comes in many varieties, and it is still gaining strength among groups and organizations. The coherence and diffuseness of the different actors and components that are determining e-research infrastructures, however, are located on different levels: in terms of e-research policy, a common agenda is driven to a large extent by national funding councils, an agenda that currently promotes openness wherever possible. Openness is also widely promoted as a worldview and institutional characteristic of science. The organizational movement advocating openness more widely is proliferating but also diffuse. Technical e-infrastructures meanwhile are consolidating and expanding into a more robust and congealed physical electronic network in which all constituent parts are part of a larger whole, including, as we have seen in previous chapters, in the production and "consumption" of knowledge. The extension of this large technological system's open capabilities depends as much on organizational capabilities as on technical ones.

Nevertheless, there continue to be tensions over the extent to which e-research and its infrastructures will be "open" or "proprietary."[19] As we have seen, this is not strictly an either–or scenario, nor, as we have seen, is the attention space or the impact of these new tools and data a zero-sum game, though both are limited. The movement promoting openness is bound to succeed in parts because of its existing momentum, but however much it gains in extending the bounds of openness, there will also continue to be closed parts, such as proprietary data, restricted publications, and networks with limited access. In terms of e-infrastructure, the walled garden of proprietary access and use and of exclusionary costs and standards will remain standing on one side, and completely open access, gratis use, and open standards will beckon on the other—and e-research will be dominated by one or the other, with some parts of functional overlap. Because much of this struggle concerns resources, the outcome is unlikely to be a clear victory for one side or the other. Instead, the scope of openness or "closedness" will be extended in particular directions, not just in reach or extent, but also in depth (the degree of nonrestrictiveness).

Before discussing the balance between open and closed systems, it is worth spelling out an important contradiction (which has not so far, perhaps understandably, received much attention): Why should national funding bodies fund e-research and an e-infrastructure to secure *national* competitive advantage when an open infrastructure will benefit the world at large?[20] Similarly, why should a university or a private company's research group make its data or findings freely available rather than securing economic gain through intellectual property rights—or at least by exploiting this resource to gain academic or domain status for itself before releasing a (less valuable) resource? This tension only needs to be stated baldly for us to recognize the force that works against it: first, that the major gain in academia or innovative research (including tools and data) in these cases may come from attention rather than from economic gain; and, second, that building the capacity or the expertise to develop these findings or data or tools within a group is itself part of the gain, including (again) the attention received by this work, which secures further talent or resources or both.[21]

The account given here so far has attempted to be neutral, pointing to the various social forces at work. Normative arguments (Benkler 2006) and pragmatic policy arguments (David 2004) can also be raised, such as the effect on overall economic growth that can be made in favor of open digital infrastructures. In this respect, it is important not to equate openness with ideas on the political left. Yochai Benkler's arguments, for example, are based primarily on liberal individualism, though some of them are social democratic (2006, 301–310). And the open-source movement is to some extent informed by a libertarian "hacker ethic" (Weber 2004). Further, openness, as we have seen, is not necessarily opposed to commercial exploitation or commercial gain. And finally, open e-infrastructures are being developed globally insofar as they are part of the public Web, regardless of political or economic systems.[22]

The normative ideas supporting openness, in other words, are varied. At the same time, the thrust of "open science," including its origin in the

Mertonian norms of "communism" and "universalism," resonates with ideas about a globally shared e-research infrastructure. Nevertheless, these normative or policy-driven ideas need to be put into sociological context: namely, how large technological systems and infrastructures develop (T. Hughes 1994) and how they are shaped by the social environment of scientific research—including different social groups and their advocacy and the receptiveness of public opinion. The chances that "openness" as an ideal and as policy and practice will succeed depends on advocacy among the research community and other social groups within a wider social environment as well as on the technological momentum of the system itself and its extension in a more—or less—open direction.

As mentioned earlier, the various parts of openness, e-research, and e-infrastructures cohere or connect only in overlapping agendas and intersecting organizations as well as in the dynamic of a large technological system. A sociologically realistic assessment of the prospects of openness can thus be made in relation to several arguments that have been elaborated here and that are interconnected. First, despite their momentum, open e-research infrastructures cannot be assumed to be extensible without taking into account their limits, foremost of which is how they compete on the research frontier for limited attention. Second, research policy can attempt to steer infrastructures toward greater openness, but this will also require greater organizational coherence among the various actors and groups promoting it more broadly as well as early support for open solutions in overcoming technical and social bottlenecks that will set the direction of the system on a particular track. Third, research policy will need to become aligned with and attempt to steer these wider forces. And finally, an open technical infrastructure and openness as a research ideal are mutually constitutive; one without the other will narrow the social benefits and constrain knowledge production because of the way in which technical instruments and data are inextricably linked to scientific advance and the advance of knowledge.

9   Limits of e-Research

The previous chapter highlighted a number of limits of e-research in terms of openness in its various senses. In this chapter, we focus on different limits, limits that are a product of various challenges in the pursuit of e-research. For some projects, funding ran out; anticipated levels of use failed to materialize; technology projects lacked the follow-on funding necessary to become infrastructures; project personnel moved to new projects and new institutions, leaving their previous projects to whither; and unanticipated barriers were raised to adoption. This chapter does not provide a comprehensive list of projects known to have succeeded but then withered or failed altogether. Also, our purpose is not to point fingers of blame but instead to illustrate some of the problems that arise with the creation of new modes of engagement and new models of working and sharing.

When Technology and Regulations Collide: GeoVue

GeoVue[1] was a node of the United Kingdom's NCeSS from 2006 to 2011.[2] The project was based at the University College London's Centre for Advanced Spatial Analysis, and its central focus was making tools that allow the fast and easy creation of maps that include data elements for visualizing geospatial data. One example of the node's output was the Virtual London demonstrator, which offered users a virtual fly-through of London using the Google Earth viewer as a platform.

When GeoVue created Virtual London, the target audience it had in mind was primarily urban planners and policy makers. Virtual London was designed not only to create a useful three-dimensional map of London but also to allow overlaying data onto the maps. For example, local governments could overlay pollution readings and visualize areas where pollution abatement measures were most needed. Once the application was completed, the project attracted the attention of Google, which expressed an interest in including it in the Google Earth viewer so that users anywhere could have access to Virtual London.

It is at this point in the story that institutional and legal arrangements intervened and prevented the public release of Virtual London. Although the Virtual London application was built in the Google Earth viewer, it relied on the Ordnance Survey's *MasterMap* data to build the three-dimensional projections of buildings. Ordnance Survey is the UK national agency "responsible for the official, definitive surveying and topographic mapping of Great Britain" (Ordnance Survey 2008). In order to include Virtual London in Google Earth, Google had to obtain permission to use the Ordnance Survey data underlying the application. Ordnance Survey data, however, is protected by Crown copyright, which covers all works "made by an officer or servant of the Crown in the course of his duties."[3] Ordnance Survey data must be licensed for use. This is in sharp contrast to the situation in the United States, where "copyright protection … is not available for any work of the United States Government."[4] Because works created by federal agencies are not subject to copyright protection in the United States, data created by such agencies can be used by individuals and by organizations wishing to put them to use.

In the case of Virtual London, Google was willing to pay for a license and indicated that the actual amount was not a great concern. According to a GeoVue staff member, "Google were willing to pay whatever it took, yeah. That was the whole point. They would pay whatever Ordnance Survey said. … [But] Ordnance Survey wouldn't budge for any price" (interviewed by the authors, October 3, 2007, London). Ordnance Survey wanted to charge £1.50 "per click," which was not possible in Google Earth, where this type of clicking is not part of the interface. In Google Earth, the user zooms and flies (virtually) through the space. Neither Ordnance Survey nor Google was able to suggest an equitable way of measuring the uses of Virtual London. Even though later discussions took place among GeoVue, Google, and Ordnance Survey, in January 2008 a member of the GeoVue project team indicated that he had given up on getting Virtual London released and had moved on to other projects (personal conversation, January 16, 2008, London). Throughout the project's duration, the *Guardian* newspaper[5] led a high-profile campaign to open access to the Ordnance Survey and other government data, a campaign that had many successes and was still active as of the end of 2012, writing in its blog in 2011, "We asked about a year ago whether we could declare the campaign done, finished, over. And the answer was clearly no, we can't" (*Free Our Data* 2011).

In other words, the issue of Ordnance Survey access was not limited to the GeoVue project, but part of a wider ongoing debate in the United Kingdom. It can be mentioned in passing that the lack of openness here also illustrates the tensions discussed in the previous chapter because the UK government has in recent years promoted a variety of initiatives aimed at encouraging "open government data" and the like.

This case illustrates how legal and institutional issues can erect barriers to sharing data and implementing innovative e-research projects. It is often assumed that many projects fail to achieve their goals because of technical barriers to implementation. In the case of GeoVue, however, the GeoVue team had successfully removed the technical barriers and had built a functioning, effective application with the potential to have an impact on public policy and academic research. Legal arrangements, in particular national copyright requirements, however, prevented Virtual London from being released widely.

Institutional arrangements also came into play. When GeoVue and Google approached Ordnance Survey for permission to use the Ordnance Survey data employed to generate Virtual London, they ran into considerable institutional indifference above and beyond the legal barriers. In other words, GeoVue not only saw the copyright and licensing issue but also had a sense that Ordnance Survey just did not consider such an application of any interest. This became even clearer to GeoVue project members after they set up a sample Virtual London in their space on Second Nature Island in the *Second Life* virtual world. GeoVue felt that the *Second Life* Virtual London demonstrator would be seen as a feather in Ordnance Survey's cap, which was clearly identified as a source of the data. The data were not exposed in the demonstrator, so the demonstrator was not making it possible for nonlicensed users to obtain protected data, and GeoVue felt Ordnance Survey could use *Second Life* Virtual London to show people how its data could be used. They were taken aback, then, when the Ordnance Survey response was a demand to remove the demonstrator immediately or risk legal action. The GeoVue project members had no choice but to remove it and replace it with another demonstrator, Virtual Phuket.

In this case, several institutional and disciplinary contexts came into conflict. GeoVue comprised academic geographers and computer programmers, for whom, as for most other academics, the notion of sharing one's results and publishing one's output without expectation of direct financial reward is part of the institutional and cultural expectation. The researchers may also have been somewhat naive about the legal issues involved because the team did not include lawyers or copyright experts. There would have been little reason to expect a need for such experts when the initial plan was to figure out ways to combine computer programming power with questions of interest to geographers, government planners, and other academics. This lack of legal expertise, however, resulted in being caught unaware when the team came up against the government bureaucrats and lawyers working for Ordnance Survey. Google's institutional culture as an American corporation also had an effect: Google's response—to assume that throwing enough money at the problem would cause it to go away—was fairly typical of the stereotype of Americans.

Although this case illustrates a constraint imposed by legal and institutional considerations, it nevertheless also shows some of the promise of transformation that projects such as GeoVue offer to social scientists. In the past, the use of geospatial information has been relatively limited because of the technical complexity of working with map data and geographic information systems programs. Recently, however, and especially with the advent of the Web, there has been a surge of interest in digital mapping techniques and combining them with different types of data and information. In the past, many social scientists who wanted to include geospatial analysis and data presentation often decided to do without this visualization component in their research because it presented too great a hurdle unless they had a geographic information systems expert on staff or available through their organization. Today, these tools and data have become more widely available than in the past, such that many projects using online maps and related visualizations are flourishing.

Finally, therefore, we can return to our larger themes. GeoVue has made high-profile contributions to e–social science in the United Kingdom, so the legal blockage to one sample of the group's work has not stopped its research: GeoVue (including its successor project) and the Centre for Advanced Spatial Analysis group have produced several other innovative research projects. Further, as mentioned, there have been many other research advances in digital geospatial mapping elsewhere. In short, this was a bump in a road where the traffic of knowledge advancement is nevertheless on the whole proceeding smoothly.

Digitally Dusty Web Archives

Another example that illustrates the challenges faced when creating research tools and infrastructures that span the social sciences and the humanities is the area of web archives.[6] In a series of projects[7] on the topic of how social scientists use web archives in their research, we found time and again that although there are pockets of use, social scientists and others have largely ignored web archives as a source of data. In 2011, we suggested four possible future scenarios—which we labeled the "nirvana" of widespread use of standardized web archives with powerful analysis tools; the "apocalypse" of fragmented nonstandard, difficult-to-use archives; the development of a "singularity" that supplants the Internet with a new (possibly intelligent) form; and the "dusty archives" scenario in which carefully collected archives of web pages sit largely unused while gathering digital dust (see Meyer, Thomas, and Schroeder 2011). The last scenario, whereby web archives remain largely unused, appears the most likely based on current trends. This conclusion was based in part on two earlier reports from our team funded by JISC,[8] which focused on the state of the art of web archives (Dougherty, Meyer, Madsen, et al. 2010) and on opportunities for new investment (Thomas, Meyer, Dougherty, et al. 2010). One of the themes throughout that work, which relied on expert interviews with the web archiving community, was that "there is still a gap between the potential community of researchers who have good reason to engage with creating, using, analyzing and sharing web archives, and the actual (generally still small) community of researchers currently doing so" (Dougherty, Meyer, Madsen, et al. 2010, 5).

In concluding that the "dusty archive" scenario was the most likely—unless serious changes were made to the way the web archiving community built collections and created tools—we argued that web archives are likely to become the digital equivalent of the dusty archive, often well curated and maintained but hardly used.[9] In this scenario, even though the web archiving community might continue to develop standards and practices for preserving portions of the Internet, few really impressive uses seem likely to emerge from the research community. Pages may be individually consulted via online tools such as the Internet Archive's Wayback Machine,[10] and some researchers will continue to build small archives for particular research topics, but Internet research will continue to focus primarily on the live Web, and little interest will develop in using the past Web for serious research in the near future (Meyer, Thomas, and Schroeder 2011, 7).

That scenario is different from one in which web archiving technology cannot keep pace with technological changes on the Internet. Here, web archiving does keep pace with web delivery technology, but the data preserved remain just that—specimens preserved for uncertain future use by archivists, but not *for* researchers. One of the reasons for lack of researcher engagement is that instead of consulting web archives, the live Web itself is increasingly seen by users and researchers *as* the archive. The live Web continues to grow, and, for the most part, many tolerate the disappearance of data as a simple inconvenience, outweighed in general by the otherwise huge volume of data that remain on the Web at any given time.

When most of us imagine an archive, we see an image of physical items such as papers and documents stored in boxes or drawers in a physical place. However, this is not the essence of what the Web is. Researchers can capture large amounts of material from different live web sources to achieve their research aims. We perceive of archives as things that are locked away for posterity, yet the Web itself is an ongoing, growing, massive, and diverse source of different types of materials that are of potential interest to researchers, which they see not as a traditional archive, but simply as a data source.

This is a pessimistic scenario, but one that currently has the weight of evidence on its side. In our consultations with a number of leading researchers, we came across a persistent lack of interest in asking questions about the past Web and in understanding the Internet as a historical development. There are of course exceptions (see, for instance, a recent paper from our research group, Hale, Yasseri, Cowls, et al. 2014), but we have not detected a latent desire for working with web archives that is simply awaiting suitable technology to awaken it. Perhaps such a change is waiting around the corner, ready to cause a step-change in researcher's imaginations based on the demonstration of a new use or a new technology. If there is not, web archives can be expected to continue to gather digital dust. We might avoid this scenario if new types of archivist could emerge—ones who engage with researchers and the public in extracting the data they need from the live Web and who, when data have disappeared from the live Web, are able to restore them in a way that makes them visible and usable to the tools of the live Web. Much as the digitization of historical documents from physical archives has made huge amounts of historical material available on the Web over the past decade (Tanner 2010; Meyer 2011; Tanner and Deegan 2011), web archives do not need to be moved off the Web into boxes, but to be moved back onto the web when the content that they contain has otherwise disappeared (Meyer, Thomas, and Schroeder 2011).

What this case shows us is how technology that does not have a clear community of researchers at the ready, as it were, stands much less chance of widespread adoption than tools that fulfill unmet needs for already active communities. Recall, for instance, VOSON and NodeXL discussed earlier: these tools are being adopted by researchers interested in SNA, but the SNA community long predates the creation of these tools.[11] Without such a community, web archives (and other collections for e-research without clear uses) will continue to struggle to find researchers who want to use them. Finally, perhaps the limit in this case is simply a matter of time and imagination: time before the Web is regarded as a valuable resource of social research about the online past in its own right (similar perhaps to newspaper archives, which were originally in relatively difficult-to-access microfiche but are now available as more widely accessible digital archives) and imagination insofar as one requirement for the advance is to recognize the value of certain objects for advancing (social) science and knowledge. It can be mentioned here that other such new objects—including *Wikipedia* and Google search behavior, but also Twitter and Facebook—have not experienced this failure, even if these new objects of research have limitations of their own.

The Limits of Data Sharing

A major issue standing in the way of widespread data sharing is the unresolved issue of how academic researchers can be assigned credit and rewards for contributing scientific and other data to a public archive. We discussed aspects of data sharing in chapter 5 but can now turn briefly to some limits of distributed data. One such limitation is the reward structure of academia, which currently favors publication in top peer-reviewed journals and the subsequent citation of one's work in other peer-reviewed articles. Most measures of academic quality offer no credit for having created a data set that other researchers use or for creating a software package that enables research. Indexing services such as the Web of Knowledge do not index databases. The promise of e-research in this area represents a potentially major transformation in scholarly practice: Should the publication of scientific data become a standard part of the scientometric measures that measure scientific impact and are often used in the evaluation of researchers, departments, and organizations (Borgman 2007)?

Some efforts are being made to address the issue of academic credit for data sharing. In the GAIN case described earlier, we saw how a public–private partnership successfully offered academic researchers high-quality genotyping an order of magnitude more detailed than they had previously used in exchange for releasing their data and blood samples for use by other researchers and private organizations such as pharmaceutical companies. The data distribution for GAIN was implemented in a system called dbGaP by the National Center for Biotechnology Information, a national resource established by the US National Institutes of Health. Among the features of the center's data-distribution site are ways of seeing the number of publications and the authorized requests made for data. Again, we see a potential for transformation in scientific practice here, one that is closer to realization through the combination of technological enablers (the dbGaP website), scholarly incentives (access to additional data), and economic incentives (access to additional future funding for scientific work).

Ethical Issues

Data sharing has proven to be a significant challenge for many efforts at building collaborative infrastructures for research. Beyond legal and institutional barriers to sharing data, such as the GeoVue case described earlier, a variety of other issues have arisen as researchers contemplate sharing data. Annamaria Carusi and Marina Jirotka (2009), for instance, have published work discussing how a seemingly simple data archive can become an "ethical labyrinth." Digital archives, they argue, can help normalize scientific research by increasing accessibility, uniformity, and transparency. Archived data can also help scientists avoid needless repetition and enable secondary uses of data. However, in practice, much social science data resist archiving—in particular qualitative research data, in which issues such as guarding subject privacy through anonymization can render the data essentially meaningless as important relationships are stripped from them. Other contextual information may not be apparent in the data, and secondary researchers may mischaracterize data in ways that the original researchers who were embedded in the social relationships under study would not. Reusing data for purposes beyond those initially stated not only has implications for informed consent agreements but also runs the risk of jeopardizing the relationship between the original researcher and their subjects if the subjects feel misled or duped by the subsequent uses.

Privacy and Trust

Privacy is also a serious concern to all researchers collecting data on human subjects. Although quantitative research data is often relatively easy to anonymize, and many quantitative researchers routinely create deidentified or limited data sets, qualitative researchers have not generally had to deal with this issue up to this point. Because few qualitative data sets have been shared outside the original research team, there has been little reason to deidentify the data. The challenge in deidentifying qualitative data is nontrivial compared to the relative ease of simply excluding several columns of quantitative data. Video recordings include images of participants; audio recordings include the sound of a person's voice; and transcribed interviews often include references to other people by name and other details. It has been suggested that improving the quality of qualitative data is an important goal for social researchers and that transparency about the data and the research process are key elements to doing so (Brower 2000). The question of how to share such rich data while continuing to protect the identity of research subjects, however, is a major barrier to collaboratively sharing qualitative research data. Whether the promise of gains offered by collaboratively sharing qualitative research data is great enough to overcome these constraints remains to be seen; if it is, it would potentially mark a major transformation in the generally solitary and opaque manner in which much qualitative data is collected.

The issue of sharing qualitative data also relates to another key issue: trust. Trust can take many forms, such as the trust people place in artifacts and the trust people have in those who make use of artifacts (Carusi, Jirotka, and Parker 2006). Trust is central to human relationships in general as well as to scientific and research collaboration in particular; without trust, people are unlikely to share and collaborate. This is an issue for both individual researchers and research participants; on a wider level, it is also a question about people's trust in government and researchers generally. Even in Sweden, which is generally a "high-trust" country in this respect (as discussed in chapter 5), maintaining this trust and extending it in the face of new e-research possibilities that are being developed take considerable effort (Axelsson and Schroeder 2009). Therefore, although it is possible to build information systems that inspire trust because of high levels of perceived security, trust can also be easily destroyed by minor breaches. Trust between individual scientists or researchers is equally hard to maintain because it relies mainly on interpersonal characteristics, though in certain disciplines (high-energy physics) the norms created in collaborations by highly bureaucratic organizations can overcome personal idiosyncracies (Shrum, Genuth, and Chompalov 2007). The challenge for e-research, then, is to understand which elements of the research process are most reliant on trust and to ensure that trust is not undermined by the technologies and processes expected of researchers.

Economic Constraints

Another concern expressed in e-research projects pertains to the economic constraints academic researchers face when compared to industry players.

Many of the funded projects in the early years of the e-research programs discussed in chapter 3 (particularly in the area of e–social science) were generally small demonstrator projects with little or no support for the eventual development of mature research tools with the extensive user support that would be required if usage were to become widespread. There was a sense among some in the e-research community that the most likely outcome of many of these academic demonstrators would be that industry would see that some of them are viable approaches to a problem, and the industry developers could then create a package of software tools or materials that would trickle back down to the research community in several years time (Peter Halfpenny, personal communication, 2008).

There was evidence for such a development, for example, at the beginning of 2008, when speculation in *Wired Magazine*'s blog reported rumors that Google was developing a project code-named Palimpset to host open-source science data in a tool that would include commenting and annotating features (Madrigal 2008b). Google's resources are far more extensive than those of academic projects, but in this case Google quietly pulled the plug on the project without releasing it by the end of 2008 (Madrigal 2008a). Should well-funded industry actors decide to move into the e-research arena in a serious way (they of course already have, as with Google Scholar and the academic networking sites Academia.edu and ResearchGate [see Nentwich and König 2012]), the entire landscape of e-research would be transformed. One area where this may happen in particular is big data, where industry players are already very active and are leading the way in developing tools—often very expensive tools—for data analytics.

Many other constraints have limited the early uptake of e-research, but these few examples indicate the main issues and potentials for transformation in science and research if these constraints can be overcome. Finally, and, again, we return to this point in the concluding chapter, these limits occur in practice: science and knowledge are in principle open and open-ended, as we have seen, and one of the roles of analyzing them from a social science perspective is to point to their shifting shapes, including barriers and ways to overcome them, as well as how these shifting shapes relate to society at large.

10 Knowledge Machines

Styles of Science in e-Research

We have given a wide range of examples of e-research efforts: those that are large and small scale, complex and relatively simple; those that organize people (researchers) and machines (research technologies); and those whose technologies manipulate research materials (data) in different ways.[1] We can now consider some common characteristics of how e-research has transformed knowledge and research practices. In doing this, we develop some ideas in the sociology of science and technology and about how knowledge relates to society more broadly. Put simply, what is distinctive in e-research is how computing performs operations on digital materials (data): e-research yokes the power of networked computing so that various operations can be scaled up, intensified, and organized in more powerful ways. To analyze these modes of research at a more abstract level, we have drawn (particularly in chapter 2) on Ian Hacking's (2002) "styles of science"—though, as we have seen, e-research typically combines several styles, and several other components—research technologies, research fronts, and communities oriented to technologies and focused on objects, plus various organizational dynamics promoting CMs and particular SIMs—are needed for a complete account of how e-research transforms knowledge.

How do all these elements advance scientific and research practice? Here we can briefly recall the argument in chapter 2: it is necessary to make a distinction between science (representing and intervening [Hacking 1983]) and how technologies (refining and manipulating [Schroeder 2007b])—here specifically research technologies—support science. As argued in chapter 2, Hacking's idea (based on Crombie 1994) that there are several "styles" of science, based on a realist and pragmatist understanding of science and knowledge, must be complemented by an account of the role of research technologies and how they enable a "mathematization" (or, here, computational approach) of these styles. Several of these digital transformations are exemplified by the kinds of e-research described in this book: Hacking's "experimental exploration and measurement of more complex observable relations" (physics and the LHC), "ordering of variety by comparison and taxonomy" (SPLASH), and "the statistical analysis of regularities of populations and the calculus of probabilities" (GAIN) (2009, 10). Importantly, for Hacking, styles are "self-authenticating" (2009, 46–47); that is, they are validated by the style of reasoning that is used.

In figure 10.1,[2] we present a schematic version of six of the cases discussed in the book to illustrate how various styles of science are enacted and how various structural arrangements support the flow of data and information across the cases. In other words, the schematic shows both the scientific and socio-organizational aspects of e-research. In the case of SPLASH, for example, individuals and teams of scientists and researchers create collections of digital materials for which the input are whales (indirectly, via photographs taken of them). Using the taxonomic style to match and sort the digital photographs of the whales, the SPLASH researchers build shared databases, which they then use to answer the scientific questions they have posed.

Pynchon Wiki follows a different model than the other cases (and recall, this is not "science" in the sense that it does not relate to material in the physical or social worlds—here we have knowledge advance in the sense of an "interpretative" taxonomy, arguably a scientization of literary research): rather than raw data created by measurement or sensing, the "data" in this case consist of the details of a novel created by a single person and the facts about the world that a skilled sleuth can match up to the author's words. Using wiki tools, the individual contributors are analyzing the original text to create annotations, which then themselves become (in combination with the annotations from all other contributors) the digital "data" (text interpretations) that others can access, share, and analyze.

The original digital data in the Galaxy Zoo case is gathered by means of the experimental style (which, remember, consists of observation and measurement, images created via the observation platforms of telescopes), and the volunteers then help perform taxonomic work on these data, sorting it into various buckets. Once these data are categorized, researchers can then analyze them and ask questions of them. In several cases, such as Hanny's Voorwerp or the discovery of Green Pea–type galaxies, the analysis by these volunteers has led directly to new discoveries and moved them

Knowledge Machines

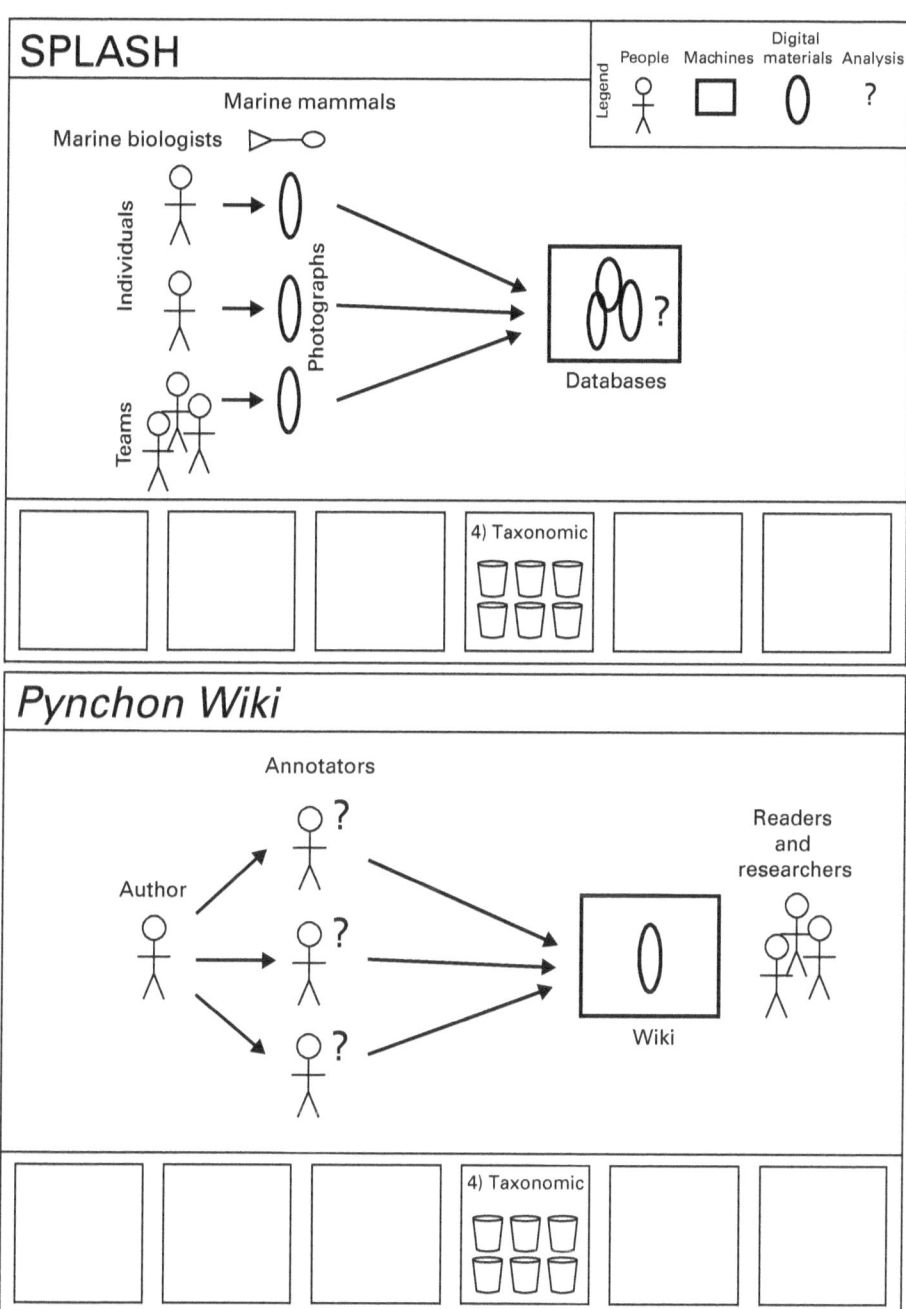

Figure 10.1 (a–c)
Styles and e-research.

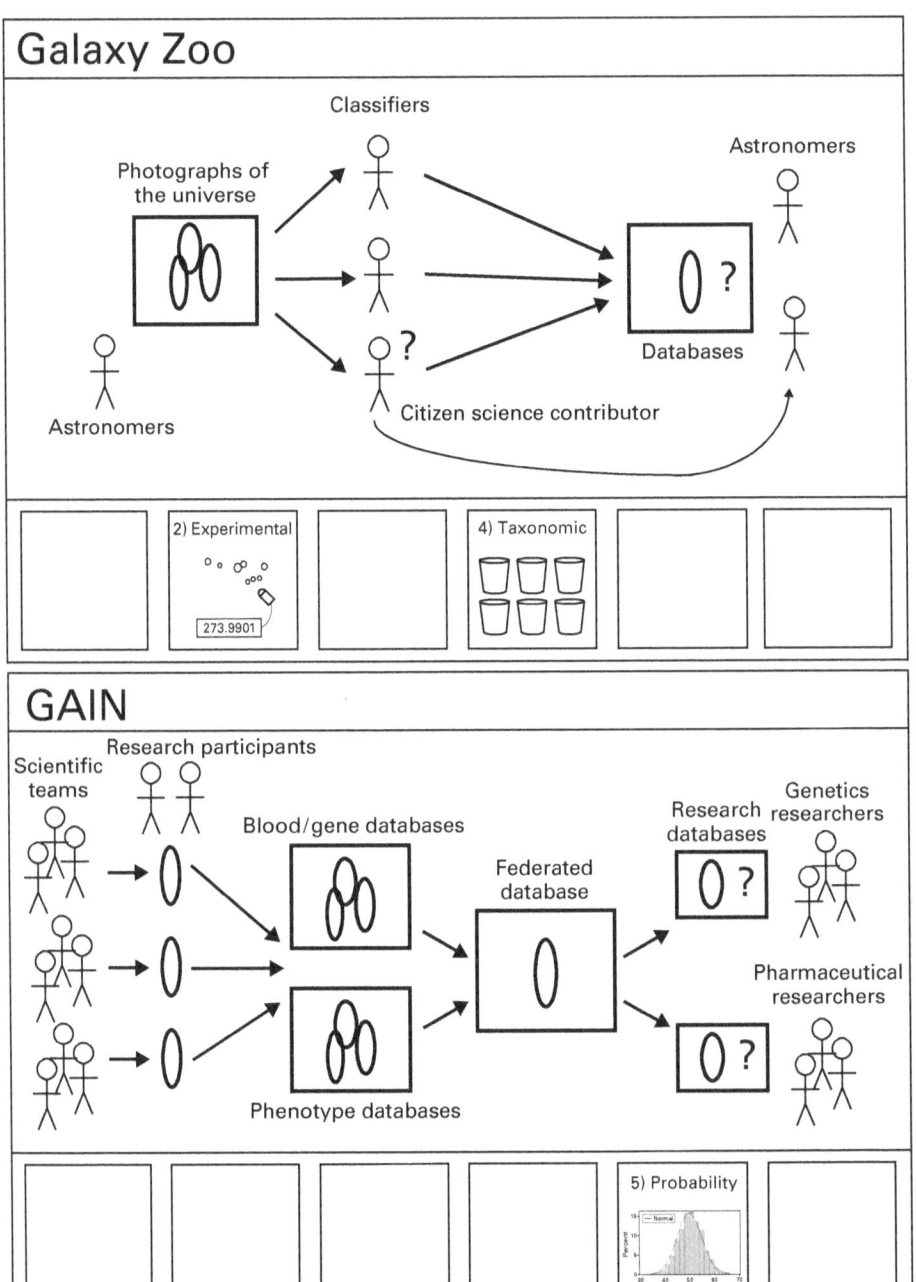

Figure 10.1 (a–c)
(continued)

Knowledge Machines

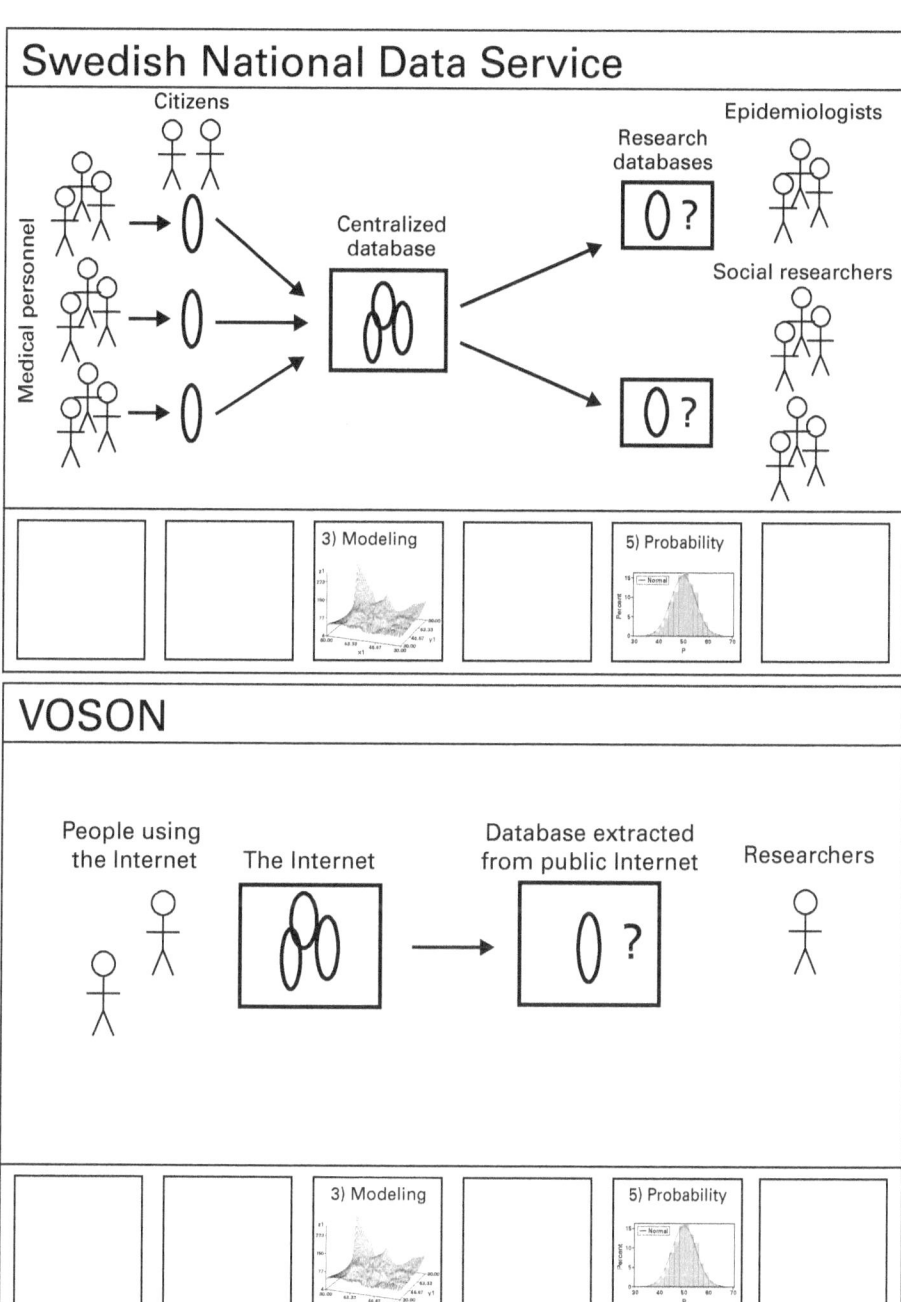

Figure 10.1 (a–c)
(continued)

into a dual role as contributors to science and knowledge directly and as typical volunteers.

The GAIN example, in which the inputs to the digital data sets are people (again indirectly in the form of blood samples, which are transformed into digital genotype data, and answers to survey questions, which are turned into digital phenotype data), the primary scientific style is statistics. This style allows the analysis of shared data sets, which, in turn, are recombined and analyzed with other data by genetics researchers, pharmaceutical companies, and other researchers. In the Swedish National Data Service case, people are similarly the inputs via the records of their interactions with governmental and health agencies. These data are combined by various actors using statistics, and on this basis models can be created that affect both research (including both epidemiology and social research) and policy.

Finally, looking at VOSON, the data (which are created largely by people's activities but are also generated by automated routines) come from the Internet. Using statistical techniques and modeling techniques (which are implemented in the VOSON and NodeXL tools), new simplified data sets are created that researchers can query in their search for answers to questions about the structure of interactions on the Internet.

Such models can be made for any of the examples, large and small, given in the previous chapters. From the experiments at the LHC, large amounts of data are being produced that need to be analyzed in a computationally intensive way, and the task is being distributed to many machines. In the case of SwissBioGrid, similarly, the task—screening molecules—is organized so that it can be performed many times on many networked computers (parallelization). Or in the case of the CENS network, streams of data from different remote sources are fed into a single store in order to collect and detect patterns within and between these data.

In certain cases, it is more difficult to speak of Hacking's styles of *science* because these e-research efforts do not (necessarily) involve the manipulation of digital data in the service of science per se: the e-research projects in the humanities are examples of human interpretation of digital materials (not of scientific "representing and intervening" in relation to the physical world). So *Pynchon Wiki*, for example, does not "intervene" in the physical world but provides a collection of materials for interpretation that are organized by technology for easy access. It is e-research, but not e-science, though technology (or research technology as per chapter 2) plays a supporting role. Such a role can be seen in all of the examples given here: distributed access to anyone with a PC and an Internet connection. So it is not always the case that e-research can be characterized as science, which tends to harness computers to manipulate data (about the physical world), but e-research in the humanities nevertheless consists of harnessing people to a common task in annotating or working on texts or a host of other materials (data in a broader sense). Even in this supporting role, however, the transformation of scholarship through technology leads to new directions in research, with the community of researchers orienting themselves to a new research object, and there is a quantitative or computational approach to this object, which is arguably a scientization of the social sciences and humanities, areas that had previously not used such scientific approaches.

Without wishing to wade back into the science wars of the 1990s, we want to discuss briefly one key point that helps to understand a major difference between science and the social sciences and humanities: science has mechanisms to reach closure (at least temporarily) on major questions, whereas the social sciences often do not, and the humanities largely do not. As Agnes Heller has argued, "The natural sciences, despite changes of paradigm, proved to be *essentially cumulative*, whereas the social sciences, despite the tendency to build up certain kinds of knowledge, proved to be *essentially noncumulative* although they claimed to be all-embracingly cumulative" (1989, 293, emphasis in original). What this means in practice is that once scientific problems are solved to the satisfaction of the scientific community, they can be taken for granted and do not require constant revisiting except to improve upon and extend earlier work. Or, as Randall Collins has argued (1994, 1998), and as discussed in chapter 2, "High-consensus rapid-discovery science" masters certain objects or domains and then moves on. Thus, later scientists will build upon the current answers to those problems and address new problems. This same point can be made in terms of a contrast with Thomas Kuhn (1962): Kuhn argued that periods of "normal science" predominate in how science works and that only when sufficient evidence accumulates to suggest that the underlying theories are inadequate to deal with increasing anomalies in scientific data will there be what he called a paradigm shift, such as that from Newtonian physics to general relativity. Kuhn's ideas, however, confront the problem of being relativist (How is one paradigm more valid and powerful than the previous one?). Hence, we have adopted a perspective that moves beyond Kuhn with Collins's ideas about cumulation and Hacking's "realism" and pragmatism, have provided a sociological grounding for Hacking, and have extended Collins's ideas about research technologies to the contemporary research landscape (Schroeder 2007b).

The social sciences and humanities are not generally cumulative in the same way that the sciences are. A landmark study that shows how Facebook networks grow will influence later related work (and be cited in subsequent literature reviews), but it will not preclude further similar studies (on new samples, in different time periods, focusing on different variables, and so forth). Instead, it may inspire later researchers to use similar methods to create new competing interpretations of Facebook networks. And with respect to Facebook as a digital object, more digital data and more powerful digital tools about this object will continue to produce novel social scientific advances, even if these advances are also limited by the object (Schroeder 2013b). A different example from the humanities might be that the definitive biography of Lyndon Johnson (say, by Robert Caro [2012]) does not mean that other biographies will not be written, and in this case there may also be different interpretations of his life parallel to and competing with this biography. The interpretive nature of parts of the social sciences and the humanities entails that certain questions are not settled, but that each new work adds to the collection of secondary literature that will inform later work.

One question that remains is whether increased scientization (or the mathematization entailed in computing research technologies) in the social sciences and humanities will influence a shift toward cumulative knowledge production that more closely resembles that of the sciences. The answer can be predicted, given the past history of the social sciences and humanities: some subdisciplines that adopt the tools of e-research will become increasingly empirical and positivist, whereas others will react against this trend, with researchers either using the tools of computation to advance interpretive abilities or criticizing the trend toward scientization. The same can be said of digital data, which, as we have seen, is in abundance in relation to certain objects, though in different ways (some objects relate more or less to the physical world and represent and intervene in or refine or manipulate it more powerfully; other objects relate to the social and cultural world, which is typically less amenable to representing and intervening or refining and manipulating in this way). Our point here is not to criticize interpretive social science (which itself is subject to debates regarding its scientificity; see, for example, Goertz and Mahoney 2012, which makes a strong case in support of scientific rigor in case-based research), not least because we ourselves engage in both quantitative and qualitative research (and have done so in this book). It is simply to lay out for view some differences between modes of contributing to the world's store of knowledge that either result in (scientific) cumulation or entail more plural and ever-shifting insights.

Finally, we can therefore pinpoint the specific role played by research technologies in e-research because e-research contributes to scientific knowledge in different ways in the sense of Hacking's styles (the data for the LHC experiment, to recall one of Hacking's styles, is produced by another technology, the collider, or the data sets in GAIN are gathered from patients, and so on) by providing more powerful research technologies and what they do (the ability to analyze the data in both cases). The role of research technologies, in turn, points to the recognition that more powerful instruments are often responsible for driving science and knowledge rather than the other way around (Schroeder 2008). In the case of the e-research capabilities of the LHC and GAIN, the manipulation of large volumes of digital data (the part played by the research technology) is central to driving scientific knowledge. Furthermore, we can recognize *how* technology does this: by developing a sociotechnical core that enables teams of researchers and distributed organizations to focus on an object and on the means to analyze the object, to produce and reproduce results, and to improve upon them. These research technologies, moreover, gain their importance because they can be used in different domains, acting (as discussed in previous chapters) like passports that allow passage between them (Shinn and Joerges 2002). We have seen this passage in many cases—for example, Galaxy Zoo and *Pynchon Wiki*, where research technologies go from being used for one purpose to being extended to new domains.

In short, to get at the core of how e-research contributes to knowledge, we can distinguish between e-research technologies' organizational role (giving researchers or teams a focus with tools and materials) and their technoscientific role (manipulating data more powerfully). These operations take place in the "black box" of science, which is not so mysterious in relation to the contribution of the research technology here and only requires expertise to be able to assess the contribution that is made to advancing particular fields (another "black box" that can be opened and is not mysterious). In other words, how do digital data that have been analyzed using digital tools improve on existing data or findings? This question can be answered, again, only by reference to the research community and the research objects toward which scientists or social scientists or humanities scholars are orienting themselves in each case (Gläser 2003). And whereas for some disciplines (those with a high degree of task certainty and mutual dependence [see Whitley 2000]) it is easy to ascertain whether such an advance has been made, in other disciplines it is more difficult to establish or open to interpretation or part of a contested terrain.

In any event, it is important to show not just how research practices are changing, but also, from the point of view of the sociology of science and technology, if and how these changed practices are advancing knowledge. So far, however, the sociology of science and technology has failed to meet this goal, in part because this disciplinary perspective currently rejects positions such as Hacking's "realism" in favor of constructivist actor–network theory (Restivo and Croissant 2008), in part because it pays little attention to research technologies (which are at the center of our argument), and finally because there is little by way of analysis of how the organization of science and scholarship is oriented toward a common research front and how the various organizational dynamics of these fronts—which in the case of e-research are increasingly dominated and driven by research technologies—advance parts of science and knowledge differently. And although it is not possible to establish precisely which research fronts are being advanced more and less—after all, that task falls also within the disciplines concerned, not just to the social scientific analysis of science and knowledge—what we have pointed to via our general account of e-research are the degrees of how research technologies and the data that they manipulate lead to an increasing scientization of research. Our argument can therefore be encapsulated by saying that science and technology are increasingly being driven by "knowledge machines," with ever larger domains of research subject to advancement via research technologies. Although the outcomes of this movement are still uncertain, the overall direction is clear.

Disciplinary Differences in the Transformation of Research

This way of understanding how knowledge is advancing leads us to turn to the question of the disciplinary differences in the take-up of digital technologies and the implications of these differences.[3] Such a discussion of e-research and disciplinarity must begin with the terms *interdisciplinarity* and *multidisciplinarity* because the introduction of digital technologies and materials into research is typically regarded as an opportunity for greater collaboration between disciplines. It can be noted immediately, however, that there are two ways in which this collaboration is typically framed: either disciplines relax their boundaries and work successfully across them, or they are unable to do so and remain stuck in—or are true to, depending on one's viewpoint—disciplinary silos or confines. In fact, there is at least one further possibility, which is that e-research becomes a new specialist discipline in its own right.

The traditional view of disciplines is as a hierarchy, with the "hard" natural sciences at the top (and perhaps physics, with its rigorous experimental methods, at the apex), the "softer" social sciences in the middle, and the humanities at the bottom with their greater pluralism and diversity but also weaker possibility for cumulative advance. For our purposes, another perspective on disciplines is how they maintain their boundaries, on the one hand, and encroach on each other's institutional turf, on the other (Becher and Trowler 2001; Fry and Schroeder 2010). In the case of digital technologies, the most immediate trend in this regard is how computer science and its various subdisciplines can be seen as invading all other disciplines (with considerable evidence for this invasion, as noted in chapter 3). At the same time, computer scientists often deny this invasion, claiming that they are merely supporting the other disciplines to do "better science" or "better research"; they are merely the enablers. In view of what has already been said, this view is often misleading. Even if in some cases technology plays a supporting role (for example, in providing access to many contributors), in other cases, as we have seen, it is integral to research—as when the scale or scope of computation is itself driving or enabling how data are collected or how they are analyzed or if there are more digital materials or data available for analysis or both. In other words—this point is obvious, but it is rarely made because the overall landscape of e-research has not been tackled as such—computing technologies and digital materials are increasingly pervasive as the drivers of research fronts across domains of knowledge. What is needed—and what we have tried to do in this book—is to get a sense of the range of ways in which researchers and technologies are being combined in advancing the frontiers of knowledge.

There are a number of well-known differences between disciplines, including whether researchers work in larger collaborative teams (as we have seen in the case of physics) or in specialized laboratories (as in the life sciences; see Knorr Cetina 1999 for the contrast between physics and the life sciences), that on the surface would seem to affect how e-research is organized. However, the GAIN example shows that the life sciences can also scale up and share resources between laboratories, and the SwissBioGrid example illustrates that life sciences can go beyond specialist expertise to engage with computer science. A common view of humanities scholars is that they work mainly as "lone wolves," but, again, *Pynchon Wiki* and other examples show that this view can be misleading. For digital research, we can therefore return to a question posed at various points in the book, especially in chapter 6: Are there differences in the social organizations of research around different technologies?

In answering this question (we have already shown some of this variety in figure 10.1), we find it interesting to reflect on how the very definition of e-research that we have used throughout the book rules out certain forms of digital research: for example, the researcher who makes a personal collection of digital materials (such as a humanities scholar digitizing images of rare manuscripts for analysis). If these materials remain purely the individual's own collection, without contributing to a shared resource that others can access, this collection should be disqualified as e-research based on our definition. Also, for this scholar's work to fully qualify for our definition of e-research, the work would have had to rely on the use of computing operations in the performance of research itself—and not simply build a shared collection of digital materials for remote access.

Computing operations have been used in at least some projects that use digital text material, as when the digital versions of the text allow text search, automated comparison of different text versions, word-frequency counts, and the like. Traditional humanities are transformed in this case not just by technology, but by how the technology reshapes the discipline: the interpretation of texts moves closer to the techniques of computational linguistics, arguably a form of scientization. One symptom of a discipline being transformed is when this type of scholarship becomes challenged as not being "real" humanities scholarship. Thus, for example, some scholars might argue that computational approaches to the text are not real research or only represent superficial research—the real task should be to provide richer interpretations of the texts (we saw this, for example, in chapter 6, with examples of textual analysis in literary studies).

STINs and the Momentum of Digital Research

The momentum toward knowledge machines has intellectual and technological sources, but, as we argued in chapter 2, it also has organizational sources in STINs (Kling, McKim, and King 2003; Meyer 2006), which can be conceptualized as a web of individuals (or organizations and movements) and artifacts that together constitute a system. We can now return to this conceptualization and examine the constituent parts of the e-research STIN where, as we have argued (especially in chapter 3), there are national, disciplinary, and local differences. The first questions to ask using the STIN framework are: Who are the relevant actors within the systems supporting e-research, and what are the core groups to which they belong?

Some of the actors here are obvious: funding bodies, scientists and researchers, technology developers, computer scientists, and computer application developers. Others are more unexpected: citizen scientists (or citizen researchers or volunteers), facility managers for shared resources (especially data managers), advocates of computational science from different fields, members of the public interested in accessing both the published outputs of science but also increasingly the raw data of science, and researchers formerly deeply embedded in nontechnologically mediated forms of research who discover new ways to interrogate their objects of research using digital tools and materials or data. The core groups to which these actors belong are also widely varied, ranging from professional organizations, universities and other research institutions, funding agencies and the groups they promote, scientific collaborations that connect individuals and laboratories to partners at different institutions and (sometimes) in different countries, and informal communities of interest such as the volunteer forums set up by the Galaxy Zoo team.

It is also possible to ask another core question suggested by the STIN framework: Who is excluded from e-research by design or happenstance? Because one of the goals of many e-research efforts is to enable direct access to research infrastructures and data, the old intermediaries who control access are increasingly excluded. Rather than through data processors, data can be directly accessed in raw form via download or Application Programming Interfaces, allowing e-researchers (especially in the social sciences) to transform and manipulate data directly. New intermediaries will often emerge, such as the companies who produce these digital objects. At the other end of the scientific process, the open publication of results that are directly accessible to the public via search increasingly excludes publishers, libraries, and scientists themselves, who would previously have been the main source for telling the public about scientific advances through various specialist and generalist outlets (such as science-related television broadcasts or magazines). Granted, the majority of laypersons who consume science still do so via these mediators, but the possibility of direct access is growing all the time, and increasing frequency of direct access may follow.

A third set of questions has to do with learning and adoption of e-research: What are the pressures or incentives to adopt digital research or computational approaches, and how is knowledge about how to implement these new practices obtained? In terms of pressures and incentives, we have seen evidence that major funding programs, which create considerable resource flows of money and expertise, have been a key driver toward the creation of new digital research tools and data sets and thus have enrolled many researchers and the public into the e-research CM. Other incentives have been aspirational (wanting to be seen as engaging with current computationally advanced approaches), led by science (the need to adopt novel computing tools such as Cloud or Grid computing to continue to make advances in an otherwise stalled research area) or driven by research technologies (in cases where this is demanded at the research front of a discipline). The knowledge about how to implement these technologies has been largely disseminated via peer-to-peer interactions in the cases examined here. Most researchers we interviewed or surveyed indicated that peers were the single most important source of information about how to research a new problem (despite the fact that, as we have argued, new techniques and materials are also increasingly visible—and findable—via search engines and the like).

Finally, we can ask, What architectural choice points[4] in the advance of e-research have influenced how and where the designs of the systems are put into place? Clearly, as we saw in chapter 3, the creation of various national efforts was a key choice point that contributed to a steep rise in interest and participation in e-research efforts, particularly in the first half of the first decade of the twenty-first century—even if these national "pushes" also furthered infrastructures and efforts that often spanned the globe. In the latter half of the same decade and beyond, when many of these initial efforts either ended or transitioned to new forms, we still saw growth in publications and activity related to e-research. Many of the leaders of the e-research efforts in the United Kingdom, the United States, and elsewhere to whom we spoke had a similar story: the "disappearance" of dedicated e-research efforts could be seen as a victory as the tools and data of e-research became embedded in researchers' routine activities and thus no longer required special attention. To some extent, this is a matter of uneven awareness because uptake of many of these tools and materials is still limited to early adopters in some disciplines and cannot reasonably be said to have transformed the mainstream of research for many disciplines (but the awareness is also uneven in the sense that those who have adopted these scientific practices or research technologies no longer recognize the changes that have taken place or that are still ongoing). Thus, for example, whereas the continued advance of high-energy physics is tightly coupled to continued advances in research technologies, the continued advance of many social science and humanities fields is much more loosely coupled to computational advances.

Overall, we have argued that how the sociotechnical interaction networks of e-research operate in various disciplines is complex, but not infinitely so. Although there are variations, they are often variations on a theme (as depicted in figure 10.1). Researchers, specific technologies, and types of data can be aggregated via networks in several ways, but at some level these ways can be categorized within a few broad configurations for the common purposes of advancing knowledge. And again, despite the variety of these efforts, they can also be regarded as an ongoing general scientization or attempted scientization via distributed and shared digital tools and data—in short, as knowledge machines.

CMs, SIMs, and the Networks Enabling e-Research

We have argued that shifts in funding in the United Kingdom, the United States, and elsewhere were major drivers behind the push to e-research.[5] In the United States, the shifts at the NSF toward transformative research and the growth in the cyberinfrastructure funding program attracted new proposals. Similarly, programs in the United Kingdom such as NCeSS/Digital Social Research encouraged researchers to focus their attention on developing tools and data enabling e-research. If new funds continue to be directed toward e-research development in both the United States and the United Kingdom, the sociotechnical momentum of these tools and data and the attendant transformations in the practice of knowledge are bound to continue.

Thus, the ability to potentially answer previously unanswerable scientific questions is only one, albeit powerful, driver behind the push to e-research. This is particularly true in certain scientific fields where access to massive data sets is required to obtain greater statistical power via the greater scope of the coverage of the data. In biomedical genetics research, for instance, few of the disorders for which scientists are trying to identify a genetic basis are triggered by a single gene waiting to be discovered. Instead, it is thought that many disorders are triggered by complex multiple gene interactions and may also be influenced by interactions between genes and an individual's environment. As such, many genetic studies discuss measuring overall genetic risk for developing a particular disorder rather than discovering a single gene responsible for the disorder. Because measuring genetic risk is much more complex than identifying a gene in the population, genetics researchers must rely on ever-larger data sets if they are to have any hope of discovering the genetic alleles that place subjects at risk for disease. One way to create such large data sets (as we have seen in chapter 4) is for scientists to contribute their collections of subject data and DNA to larger collections, as in GAIN. By increasing the number of DNA samples available and simultaneously increasing the amount of data available from each sample, genetics researchers hope to find better indicators of genetic risk in the populations of study.

Another type of research that can only be done collaboratively is astronomy. Astronomers have long shared resources such as telescopes and sky data. As we saw in chapters 4 and 6, several projects are currently contributing to virtual observatories: AstroGrid in the United Kingdom, NASA's national VO in the United States, euro-VO in Europe, and IVOA globally. These projects have been built to enable astronomers to work with data that is stored and distributed across the Grid and analyzed using parallel-processing techniques and shared data sets. The AstroGrid project, for instance, can find, process, and analyze data from a given patch of sky that have been collected from any instrument to which the project has access on the Grid. Only the small portions of the data that are the result of calculations need to be transferred to the astronomer rather than a large data set that he or she would have to download to work with. As these tools and databases become more widely used, the individual astronomer is able to spend less time finding and collecting data on his or her own and more time doing analysis.

A different example (related to SPLASH, discussed in chapter 5) of how collaborative resources can transform practice in a scientific field can be taken from oceanography. Robert Lamb (2006) has described how oceanographers are now able to use remote sensing to transform the practice of event-driven science. The traditional way of collecting oceanographic data was to send ships out to sea and wait for the types of events to occur for which scientists needed data. In recent years, however, oceanographers have been building extensive networks of remote sensing devices throughout the oceans. For instance, the Ocean Tracking Network[6] uses a network of fixed receivers located in the ocean to record when tagged animals pass. Now the scientists can monitor the readings from their offices. Further, if an interesting event begins about which they would like to collect additional information on site, the remote sensors notify them, and the researchers can then dispatch a ship to the location of interest. This process represents an enormous saving in time and money for the oceanographers and allows them to collect more data than before.

If the necessity of harnessing large-scale resources is a major enabler for e-research and provides this movement's network with a momentum of its own, this may offer a clue to the relatively slow uptake of the collaborative and Grid-based elements of e-research among social scientists—apart, that is, from the instances where they have begun work on the massive amounts of readily manipulable data from the digital platforms discussed in chapter 6. Social scientists have traditionally been trained as autonomous researchers, working by themselves or in small teams. The data sets they thus generate are often small and easily managed and analyzed using desktop computing applications. The exceptions here are when large-scale national data sets need to be combined and linked, which require more organizational networking rather than a scaling up of computational power. Unlike running complex astronomical calculations, even the largest social surveys rarely tax the processing capabilities of SPSS or SAS on the PC. For many social scientists, Grid or Cloud computing does not offer compelling applications, though this view is beginning to change with respect to the analysis of the data sets derived from digital platforms. Even for social scientists using quantitative statistics or large-scale SNA, their computational requirements do not generally approach those of scientists modeling complex biological, physical, or astronomical systems.

In short, a funding push, the advance of research technologies that favor certain types of research, and certain scientific questions that demand greater computational capabilities and needs for data have congealed into a CM in which the computational or algorithmic and data-intensive approaches are put forward as a preferred sociotechnical order. The flipside of this computerization are the SIMs in which new types of scientific questions and approaches are gaining preeminence in a variety of disciplines: as argued throughout, the effort to digitize scholarship can be seen as both a SIM and a CM. However, we can now add that one characteristic of SIMs and CMs is missing here: the opposition to an existing orthodoxy or school of thought. This movement or set of movements toward knowledge machines is also rather diffuse. It has been able to mobilize resources—funding, conferences, journals, and the like—under various umbrellas (cyberinfrastructure, etc.), but the only common denominator apart from the definition we have provided is that this movement has claimed to be able to enhance knowledge or scholarship. The only "opposition" here are either those who deny this claim (humanities scholars who argue that traditional approaches are superior) or the forces that shift priorities elsewhere (as when the UK funding bodies abandon their e-research programs in favor of new programs such as Digital Economy). This movement is therefore pushing at an open door, the door to advance scholarship, and on the other side of that door new directions run side by side with existing approaches or there is a lack of continued momentum.

Although digital transformations have therefore been contested or diminished only to a limited extent—and of course they are still ongoing and expanding as a movement—it is also the case that they have already in a sense "conquered" the world of scholarship: such transformations have become well established at least to some extent in all major realms of scholarship. Yet, as we have seen, these digital transformations are in part complementary, in part novel, in part additional, and overall proliferating—but not replacing or advancing beyond certain research fronts. Knowledge machines should not be seen as a revolution or a paradigm shift, but rather like ripples in a pond, fanning out across the waters of research inasmuch as scholarship has been transformed to greater and lesser extents.

Does this characterization apply also to "practices"? Here we have painted a mixed picture in the sense that we can point to a great variety of novel practices—new skills, new organizational forms, new types of outputs, and the like. With respect to practices, it is necessary to step back and make a rather reflexive point: the focus on practices has been undertaken in science and technology studies and cognate areas in order to move away from hype and toward "actual practices on the ground," typically in order to deflate the claim that radical shifts have taken place or are taking place. However, this position can be turned on its head: it is never possible to identify researchers' radically shifting practices except with hindsight (this is also why researchers themselves are likely to be unaware of them, as mentioned earlier). Instead, transformations in scholarship take place at the aggregate level, and it is these transformations that we have identified as spreading ripples. It is nevertheless true, as the science and technology studies position on practices suggests, that the novelty of these new practices, even in the aggregate, is rather mundane. But again we must take a step back: mundane, yes, compared to the changes in knowledge that have taken place, but the mundanity consists of the fact that communities of researchers are always orienting themselves toward the leading edge of the research front and competing for attention there. Slow and incremental advances are taking place in a mundane way at each of these leading edges. What is more remarkable is the change that has taken place or been common across these leading edges—namely, that digital transformations have become an essential part within and between so many of them. And finally, our approach here has been to focus on "practices," thus avoiding an "idealist" account of science and knowledge and avoiding the criticisms of a science and technology studies perspective that seeks to instantiate such accounts "on the ground." Though we have found considerable variation "on the ground," we have also identified commonalities between them.

Thus, across different areas, this digital transformation also now constitutes a research front of its own. This front consists of developing digital tools and data for the more intensive and extensive manipulation of knowledge—a "machinification" of knowledge. This does not mean that digital transformations dominate the research agenda; it simply means that uses of research technologies across various niches are a precondition for being at the leading edge (or in more pluralistic domains, among the leading edges) of the research front. This point can be related more broadly to Hacking's styles because the manipulation of data takes place in different ways in science (and also, mutatis mutandis, in the social sciences and humanities). Or we can see these research technologies as part of a system or a set of networked systems that have achieved a certain "technological momentum" (T. Hughes 1994) of their own.

Hence, the discussion about shifting disciplines via digital transformations can also be reframed: the shift is not so much a blurring of boundaries or the need for multidisciplinary skills or increasingly interdisciplinary collaborations and the like (though these are also important topics), but rather that as digital transformations take place across a wide swathe of disciplines, they entail the addition of new skills to knowledge production, either with computer scientists brought on board or when these skills and competences are developed among the researchers themselves within disciplines. In *all* cases, however, some reconfiguring (Dutton 2005) takes place.

At a macrolevel, a change toward "big science" and "large technological systems" was documented in the twentieth century, with physics the main example, which has perhaps been superseded by life science and a move toward "laboratory"-based science (Knorr Cetina 1999). Digital transformations complicate this picture in a multiform way: the situation is not merely that large infrastructures dominate or that individual researchers are tied via their networked terminals to a global body of scholarship. Rather, an ever-growing proportion of scholars and organizations now contribute to and depend on networked digital tools and data of various kinds. Thus, science and knowledge change shape again: not "big," not "large," not "lab," not "networked science," but science and scholarship advancing as part of a network of partially overlapping instruments and the data sources that they manipulate and that, even if they do not form part of a single unit, are nevertheless accessible via a single network. These networked instruments and data (or materials) are evolving more and less fluidly across various boundaries in the research landscape (though they are more pronounced in some than in others). It can be foreseen that this paradigm of networked instrumentation and databases will become dominant throughout the world of research even if it remains polymorphous. Hence, the label *knowledge machines* should also simply be regarded as shorthand. Nevertheless, this shift and its momentum are due to the requirements of ever greater complexity of research coupled with the greater need for these research technologies and forms of data, which is why it is important to begin to think both about the need for greater reflexivity about this way of producing of knowledge and why this network of machines is becoming both more autonomous from and differently coupled to society at large.

Variety and Homogeneity in Transformations of Knowledge

In this book, we have provided a wide range of illustrations of how scholarship is being transformed by digital tools and data. What are we to make of this shift? Although many transformations can be documented and analyzed among early e-research projects, a complete grasp of the more widespread transformations of research practice has remained elusive. Many projects and infrastructures have developed, and certain scientific and other fields have undergone fundamental changes in the types of questions they are asking. Yet for many scholars e-research has not become central to their work. One of the challenges in understanding the changing nature of e-research is that many of the projects developing digital tools and data have, until fairly recently, developed technologies or collected materials that have not yet been released or tested widely in real-world settings. As these efforts have matured, it has become feasible to study these projects (or larger infrastructures) and develop an understanding of the intellectual and social shaping that has occurred and of the ethical and legal issues that are associated with the projects.

At the microlevel, as observed some time ago, "the obstacles to the effectiveness of e-science are not so much technical, as social" (Schroeder and Fry 2007, 563), and overcoming these barriers can only be observed as an ongoing process. At the mesolevel, it becomes harder to identify the feedback loops in overcoming these barriers and to gauge commonalities and variety among projects and infrastructures, especially as practices are more diverse and there is a plurality of research communities oriented to diverse objects. Thus, digital transformations are going in new directions even as they also become invisible as digital tools and data become more widely used, and these new directions are furthermore affected by larger and ongoing shifts in the research environment, such as the migration away from physics to the life sciences and the greater harnessing of research to measurable impacts. At the same time, as we have documented, the tools and approaches to the manipulability of data have also transferred between various applications and disciplinary domains. In short, understanding these transformations is a moving target. Certain elements in the landscape have become well defined, such as electronically accessible journals, whereas others, such as big data for social scientists, are still in their early days. We have covered much diverse ground in this book, but it is also easier to look back at these transformations with a rearview mirror.

These points can be expressed differently: any general arguments about the transformations of e-research or attempts at a comprehensive picture have limitations and are bound to be improved upon but are at the same time essential—unless we are resigned to the particularism of specific questions and cases or to knowledge only about individual instances rather than the broader digital transformations in research. What we have shown is how these transformations can be conceptualized in a coherent way by combining a number of elements of understanding science and knowledge, how they are organized and focused on objects and research fronts, and how they congeal and gain momentum as STINs comprising organizations and groups, research instruments and data sources, and their systemic and routine relationships and interactions. These knowledge-generating STINs have involved new actors in research and excluded others, have been used to mobilize resources, and have changed shape as the result of decisions taken at various architectural choice points along the way. How these transformations are regarded is also a matter of whether one is interested more in broader changes in knowledge and scientific knowledge or in project- or discipline-specific changes or in scholarly practices and communication. Combining the specific changes in a larger picture highlights certain aspects of the transformation in question, but this larger picture is also more useful because it is informed by the many cases, examples, and themes that we have pursued. In some ways, technology—in the form of collaborative and networked digital tools and data for producing knowledge—is now tightly coupled with most (if not all) forms of research, albeit unevenly spread across the landscape of research, and even the least technology-dependent forms of knowledge production are nevertheless influenced by this shift toward a globe-spanning, computer-based knowledge machine that both generates knowledge and refines the technologies that are used in its production.

New Mechanisms

On the one hand, we have argued that it is too early have a complete understanding of this transformation because it is still "science in the making" (or research in the making). On the other, we have identified several specific and more general patterns. This identification allows us to address some of the broader implications of the transformations we have charted: Why are these changes of interest not just to sociologists of science and knowledge or to those concerned with science policy, but also to society at large and wider publics? Some of the main patterns and their implications (elaborated in part in earlier chapters) can be summarized as follows.

First, the shift to digital research generally makes knowledge more accessible. This greater accessibility has a number of implications. One implication is that academic knowledge is made available to the public not just in the public's participation in the projects or infrastructures we have discussed, but also insofar as the public can search for or come across academic materials if they are available on the open Web. Moreover, there are also implications for students and scholars outside particular disciplines, who also now find material online instead of only in their institutional libraries or among their disciplinary peers. There are as yet few studies of this shift, but we know that searching for and using online materials have in a short time become the default. Note that in addition to the increased reflexivity of research about itself, another implication is that there is bound to be more traffic across disciplines and other boundaries (institutional, national, and the like) in this new environment of general "searchability" or "findability." Note, too, that apart from the creation of new winners and losers, it is not clear what the implications are for the quality of what is found: more materials and tools can be found, but what is found is often determined by commercial search engines, and it may also become more difficult to discriminate the most authoritative and useful knowledge.

Second, as more researchers are drawn into and contribute to the digital transformations of knowledge, this part of the research landscape is increasingly dominating the attention space. Conceived of as a CM or SIM, the digital transformation of research is enrolling an increasing number of actors who aim to put the digital knowledge at the center of their research community. For example, funding bodies are creating e-infrastructures that aim to promote the use of centralized resources among research communities or user groups. This is a not necessarily a zero-sum game, whereby resources and researchers are drawn from other areas toward e-research. Nevertheless, there are implications, as we have seen, for new roles and rewards and ways of organizing research.

Third, e-research is part of a larger shift toward online knowledge in the form of blogs, wikis, online videos of lectures, and the like. Thus, it has become routine, when looking for knowledge, to expect to find it online. This raises the question whether the researchers who make materials available openly online or who participate in networked collaborative teams or who build shared infrastructures have an advantage vis-à-vis those who do not do these things. Again, this transformation is not a simple one-way shift, yet it behooves us to ask again how more direct access also requires new skills and how more complex or in any event different forms of knowledge impose new and additional demands as well as enabling greater and more direct access and participation.

Fourth, digitally transformed knowledge relates to its objects in different ways depending on the discipline or the research front within it. It is important to distinguish here between how knowledge is communicated (scholarly formal communication and informal communication), how knowledge advances, and the role of knowledge in society. Recognizing the difference allows us to tease apart some implications: knowledge advances or grows in relation to its objects, without any necessary implications for society (except indirect ones) and without necessarily having entered into the flow of scholarly communication (though it is not recognized as knowledge unless it does so). How this growth of knowledge advances and how it reshapes the physical, social, and cultural environment, including by providing the public with access, is a topic we have barely addressed here (with the exception of public access), but there are obvious implications in some areas—for example, where more effective mastery of environmental or social problems require more powerful knowledge to address and overcome them even as these problems are also being added to and constrain us (see Schroeder 2013a, chaps. 5 and 7). These implications are beyond the scope of a book about e-research except to say that as knowledge machines bestride the natural and social world on an ever greater scale and with greater scope, researchers and publics will be forced into greater awareness about how these digital tools and data work, both disintermediating between researchers and objects (for example, more and more immediate access) as well as intermediating between them with greater complexity (the ability to assess computational representations, for example, or understanding how organizations share digital data). We discuss this point further shortly, but again we can see both greater mastery and new requirements.

Fifth and finally, the external conditions of knowledge or its organization have been reshaped—for example, allowing more centralization of research in some cases and more decentralization in others. Yet the internal conditions have also been reshaped, as when the practice of research is conceived of as making knowledge digitally manipulable and subject to aggregation in online collections of materials from the start. These changes constitute a scientization or technological transformation of knowledge; a computational approach to data and the objects they represent at the research front. The wider (nonresearch) implications of these changes are that these knowledge machines occupy an ever more central place in society, with the added consequence that researchers are made more aware of how their research is perceived by other researchers (outside of their domain) and by a wider public. Part of the STIN of e-research is the attempt to automate the advance of knowledge in various ways, and the idea of making knowledge into a mainly machine-driven process has now also appeared over the horizon (Evans and Rzhetsky 2010). This machine-driven approach includes creating workflows, manipulable data sets, and the like, but it also attempts to identify patterns by computational means whereby the advance of knowledge itself can be tracked and steered. Again, this "autonomization" of science and knowledge in society is a process in which the implications summarized so far are brought together.

Thus, the various patterns that we have identified have an overall momentum that can be seen most clearly in efforts to identify broader patterns within knowledge creation and dissemination. So attempts have been made to understand the direction of knowledge by reference to which articles are being read in different fields (Van de Sompel, Payette, Erickson, et al. 2004) or which types of distributed collaboration are most successful (Cummings and Kiesler 2005) or what the trends are in research funding (Börner, Chen, and Boyack 2005). Another effort is "web science" or the semantic web, which aims to create a set of links that are queryable across domains of knowledge and practice. What these efforts have in common is that they use computational techniques to produce knowledge about knowledge creation. Yet the totality of science and knowledge also eludes capture within a single computationally analyzable sphere: there may be compendia of knowledge, such as encyclopedias or systematic reviews or data stores or other knowledge hubs, but it is not possible to query or otherwise scope all of them universally. Nevertheless, these and other efforts to shape science and knowledge can themselves be seen as a product of the digital transformation of research in its various forms and constitute part of the ongoing effort to extend it.

These efforts also consist of enrolling new actors, and we have presented evidence both of diffusion as well as of a lack of uptake and even resistance to new tools, materials, and practices. At the same time, e-research has established new niches as well as whole areas where it has become routine. We have seen that many actors—funding bodies, researchers within various fields, tool builders, and contributors of materials—have promoted e-research. All of them can be seen as part of a CM, a movement to drive research with new technologies. We have provided a number of reasons why it is difficult to gauge this transformation in midstream (and we can add, as discussed in chapter 3, that e-research often consists of tools or materials and scholarly communication that are not cited as such). Nevertheless, we have pointed to a number of instances where this CM/SIM has made significant inroads. Ultimately, the measure of success will not just be what is happening at various leading edges of the research front toward which the scholarly communities, including new ones, are oriented, but mastery of the objects of research and the moving on to new territories. It is also possible to foresee when e-research will come to an end: when digital manipulation no longer completes tasks as well as or yields as much as other, more innovative ways of pursuing research. For example, at what point is it no longer useful to enhance a digital model of climate change by adding more data (such as from new sensors)—when these new data fail to increase the usefulness of the model?

The digital transformation of research is thus not a single change, but a confluence of several partly overlapping changes. First, and most broadly, e-research changes the role of research in society insofar as it enables public participation and engagement with research. Here we can think not only of the volunteer computing projects (such as Galaxy Zoo) but also of ways in which those outside of academia use digital tools and data for research. The role of research is also changing with the increasing searchability and findability of research. Second, the organization of academic research has changed in the direction not only of greater accessibility of digital tools and materials but also toward more joint and distributed efforts. Third, there have been movements to place e-research at the center of the research front in different areas or to occupy the focus of attention within research communities. And finally, at the core of the transformation is a change in the nature of knowledge, which consists to a greater extent of digitally manipulable materials, with the added implication that knowledge becomes more reflexive or more subject to control by technologies.

These four trends overlap in some cases more than in others, but together they add up to a set of ripples that have spread far and wide throughout the world of research. There are thus a number of implications apart from those mentioned. First, for research, and in this case primarily for the sociology of science and technology, which often relies on case studies, we have shown that it is possible to show how science and knowledge are changing through a series of linked transformations. Science and technology studies needs to be broadened in order to address such wider implications for knowledge transformation.[7] Theories of CMs, as enacted by STINs, have so far been used to address organizational changes such as the adoption of information technology across disciplines, but these theories have not focused on research fronts and communities, as we have. SIMs do this, but they have the opposite problem: SIMs tackle intellectual change without paying heed to the sociotechnical systems that enable it. Studies of research organization and research policy have informed our analysis of digital transformations, but they do not address the role of science and knowledge as such or vis-à-vis the public. Public understanding of science as a field addresses the latter but does not extend to public participation in the creation of science and knowledge per se, nor does it deal with how scholarly communication is changing vis-à-vis the public. The latter topic is treated in information science, though this discipline is focused on scholarly communication rather than on transformations in knowledge and science. In short, we have argued for an expanded and interdisciplinary account of digital transformations that combines and enlarges these perspectives, provides a more encompassing understanding of this phenomenon, and is in keeping with the fluid boundaries of the phenomenon being investigated.

Second, on the substantive side, we have presented a number of reasons why we think the ripples of digital transformations will continue. This persistence entails that the implications of a number of trends we have identified will also continue: research will become more porous vis-à-vis the public and will continue to become more networked. The research front and the communities oriented toward it will take digital research ever more for granted, and science and knowledge will become more machine enabled and thus capable of greater steering via machines. One consequence is thus both the increasing autonomization of research (it becomes more embodied in machines and less in individual researchers) and the automation of research (it becomes subject to machine processes such as workflows rather than being produced ad hoc). Again, this consequence increases transparency—the mechanisms of machines and their entwining with organizations become more formalized and visible—but in other ways the production of knowledge becomes more complex, as when direct observation is displaced by the machines and the organizations in which they are embedded. In short, the very production of knowledge becomes more accessible but also more mediated and disintermediated.

There is no single way in which this process takes place across disciplines, yet there is also a commonality among many disciplines here: the increasing use of digitally manipulable materials (or data) across networks. A persistent major difference among disciplines is how these materials represent and intervene in the objects of research and to what extent the research community is oriented toward these materials. But the common focus on networked digital materials nevertheless makes it possible to think about the developments in the interaction between, on the one side, research technologies and the scholars who use them and, on the other, a wider public that seeks to understand the role of knowledge in society and that lives with the transformations these machines bring about in the wider physical, social, and cultural environments. Such an understanding can, we hope, in turn enable us to exploit these knowledge machines more usefully and powerfully in the future.

Notes

Chapter 1

1. Galaxy Zoo: http://www.galaxyzoo.org/.

2. Europe Future Internet Portal: http://www.future-internet.eu/.

3. Portions of this section are based on Meyer and Schroeder (2013).

4. EGEE: http://egee2.web.cern.ch/egee2/.

5. EGI: http://www.egi.eu/.

6. GridPP: http://www.gridpp.ac.uk/.

7. The term *Grid computing* simply refers to a network of federated computers for computation and storage of data, used in distributed mode, and answering the need for providing a common-pool resource to those researchers and disciplines requiring lots of computing power. The name draws a parallel between these computing infrastructures and the electric power grid (Foster and Kesselman 2004).

8. Galaxy Zoo: http://www.galaxyzoo.org/.

9. See papers at http://blogs.zooniverse.org/galaxyzoo/category/paper/.

10. Hanny's Voorwerp: http://www.hannysvoorwerp.com/.

11. CENS: http://research.cens.ucla.edu/.

12. Swedish National Data Service: http://www.snd.gu.se/en.

13. SwissBioGrid: http://www.swissbiogrid.org/.

14. VOSON: http://voson.anu.edu.au/.

15. *Pynchon Wiki*: http://pynchonwiki.com/.

16. ESRC Grants RES-149-25-1022/RES-149-25-1082, "Oxford e-Social Science Project, Phases I/II," Principal Investigator (PI) William Dutton; ESRC Digital Social

Research Programme Grant, "Scoping e-Research in the U.K. and the U.S. Using New Digital Methods," PI Ralph Schroeder.

17. JISC Grant, "Digital Impacts: A Synthesis Report and Workshop," PI Eric T. Meyer; JISC Grant, "Researcher Engagement with Web Archives," PI Eric T. Meyer; JISC Grant, "Usage & Impact Study of Digitised Resources Funded under the JISC Phase 1 Digitisation Programme & Toolkit for the Impact of Digitised Scholarly Resources," PI Eric T. Meyer. Note that at the time of this funding, JISC (which originally stood for Joint Information Systems Committee) was a funding agency. It has since been restructured as a registered charity and renamed Jisc (see http://jisc.ac.uk/). Throughout this book, we refer to it as JISC to recognize the organization that provided our funding.

18. Alfred P. Sloan Foundation Grant 2012-06-17, "Accessing and Using Big Data to Advance Social Science Knowledge," PI Eric T. Meyer.

19. European Commission Grant FP7-ICT-2009-5/258138, "SEServ: Socio-economic Service for European Research Projects," PI Burkhard Stiller; Grant, "The Role of e-Infrastructures in the Creation of Global Virtual Research Communities," led by Empirica, Bonn, Germany.

20. Research Information Network Grant P33, "Physical Sciences Information Practices," PI Eric T. Meyer; Grant P28, "Humanities Information Practices," PI Eric T. Meyer.

21. Indiana University grants from the Rob Kling Center for Social Informatics and the Margaret Griffin Coffin Fund, School of Library and Information Science, "Sociotechnical Perspectives on Digital Photography: Scientific Digital Photography Use by Marine Mammal Researchers," PI Eric T. Meyer.

22. Details of specific methods and sample sizes are available in papers referenced for case studies throughout the book.

Chapter 2

1. Hacking addresses this criticism with cognitive history (2012, 606–607), but this argument is philosophical rather than sociological. Chunglin Kwa (2011) provides a more thorough historical account of scientific styles.

2. Examples are also given in Schroeder (2008).

3. This is good point for terminological clarification: *knowledge machines* is a broader term than *research technologies*. Research technologies are a subset of knowledge machines, consisting of the instruments that perform manipulations of data, whereas knowledge machines also include the wider set of technologies that are used in the distributed and collaborative communication and handling of research materials in e-research, such as online databases.

4. For the American National Election Surveys, see http://www.electionstudies.org/.

5. For the British Household Panel Survey, see https://www.iser.essex.ac.uk/bhps.

6. For the World Internet Project, see http://www.worldinternetproject.net/.

7. For Europeana, see http://www.europeana.eu/.

8. For NGram Viewer, see http://books.google.com/ngrams.

9. Portions of this section are based on Meyer (2008).

Chapter 3

1. Portions of this section are based on Schroeder, den Besten, and Fry (2007).

2. NSF database: http://www.nsf.gov/awardsearch/index.jsp.

3. Scopus search term: TITLE-ABS-KEY({e-Humanities} OR ehumanities OR (humanit* W/15 digital) OR ((humanit* W/15 online) AND (humanit* W/15 comput*)) OR (humanit* W/5 infrastructure) OR (grid W/5 humanit*) OR (computational W/5 humanit*) OR {e-social science} OR {esocial science} OR {digital social research} OR ("social sci*" W/5 infrastructure) OR (grid W/5 "social sci*") OR (computational W/5 "social sci*") OR {e-Research} OR eresearch OR {e-science} OR escience OR cyberinfrastructure OR (cyber PRE/0 infrastructure) OR {cyber-infrastructure} OR {e-Infrastructure} OR einfrastructure OR (grid W/5 humanit*) OR (grid W/5 "social sci*") OR cyberscience OR {cyber-science} OR (cyber PRE/0 science) OR cyberresearch OR {cyber-research} OR (cyber PRE/0 research) OR (computational W/5 science) OR ("big data")). The data were last updated on January 15, 2013.

4. Scopus search term: TITLE-ABS-KEY(grid W/3 comput*) AND (PUBYEAR > 1992) And (PUBYEAR < 2013).

5. Scopus search term: TITLE-ABS-KEY(cloud PRE/0 comput*) AND (PUBYEAR > 1992) AND (PUBYEAR < 2013).

6. NexisUK search: All English Language News Search for ("cloud computing") (Duplicates removed).

7. Scopus search term: TITLE-ABS-KEY(big PRE/0 data*) AND (PUBYEAR > 1992) AND (PUBYEAR < 2013).

8. Scopus search term: TITLE-ABS-KEY(supercomput*) AND (PUBYEAR > 1992) And (PUBYEAR < 2013).

9. One thing to note about the humanities data is that Scopus has somewhat less broad coverage of arts and humanities publications than it does of science and social science journals. This difference is due in part to issues regarding the selection of journals for inclusion in Scopus, but it also reflects the different nature of the

humanities, where the single-author scholarly monograph is still a leading mode of publication, a mode that is not included in the journal-based indices.

10. Using a slightly simplified search term, due to interface limitations: ({e-social science} OR {esocial science} OR {digital social research} OR ("social sci*" W/5 infrastructure) OR (grid W/5 "social sci*") OR (computational W/5 "social sci*") OR {e-Research} OR eresearch OR {e-science} OR escience OR cyberinfrastructure OR (cyber PRE/0 infrastructure) OR {cyber-infrastructure} OR {e-Infrastructure} OR einfrastructure OR (grid W/5 humanit*) OR (grid W/5 "social sci*")) and not srctype(jnl or pat or sc or mdc).

11. Sustaining the EEBO-TCP Corpus in Transition (SECT): http://www.bodleian.ox.ac.uk/eebotcp/sect/.

12. EEBO: http://eebo.chadwyck.com/home.

13. EEBO-TCP: http://eebo.odl.ox.ac.uk/e/eebo/.

14. Histpop: http://www.histpop.org/.

15. The corresponding author is also the first author in the majority of cases in the data set as follows: the first author in 12,040 instances (92 percent), the last listed author in 649 instances (5 percent), and some other author in 445 instances (3 percent).

16. The standard base map for the overlay was created using closeness between various journals based on data regarding how articles cite one another from a full data set of the Web of Science. Details are available in Leydesdorff, Rafols, and Chen (2013).

Chapter 4

1. Portions of this section are based on den Besten, Thomas, and Schroeder (2009). The material for this case study emerged through participant observation and interviews, ranging from informal conversations to in-depth interviews.

2. SystemsX: http://www.systemsx.ch/.

3. Open Biological Ontologies: http://www.obofoundry.org/.

4. This case is based on work described in more detail in Meyer, Bulger, Kyriakidou-Zacharoudiou, et al. (2012).

5. Note that this is the total number of classifications, not the number of valid classifications after data processing. Cleaning the data removes approximately 4 percent of the total. For details of this processing, see Lintott, Schawinski, Slosar, et al. (2008).

6. Portions of this section are based on Schroeder and den Besten (2008).

7. Flickr Commons: http://www.flickr.com/commons.

8. Great War Archive: http://www.oucs.ox.ac.uk/ww1lit/gwa.

9. *Pynchon Wiki*: http://pynchonWiki.com/.

10. Pynchon-l: http://www.waste.org/pynchon-l/.

11. Pynchon web pages: http://www.pynchon.pomona.edu/; http://www.themodernword.com/pynchon/index.html; http://community.livejournal.com/ru_pynchon/.

12. Two others are Grant (2001) and Grant (2008); see also Mead (1989) for a bibliography of secondary works, and the ongoing bibliography in the journal *Pynchon Notes*.

13. *Thomas Pynchon* web page: http://www.hyperarts.com/pynchon/.

14. Citation statistics for both *Weisenburger* and *Pynchon Wiki* are according to Google Scholar as of January 15, 2013.

Chapter 5

1. Earlier versions of portions of this section were presented at the 2007 NCeSS conference in Ann Arbor, Michigan, and published as Meyer (2009). Portions of this work were supported by the Rob Kling Center for Social Informatics, Indiana University, Bloomington; the School of Library and Information Science, Indiana University, Bloomington; and the UK NCeSS through a grant from the UK Economic and Social Research Council (RES-149-25-1022).

2. See, for instance, the Earth Science Markup Language (Ramachandran, Graves, Conover, et al. 2004) and the Kepler system for dealing with legacy data in scientific workflows (Altintas, Berkley, Jaeger, et al. 2004).

3. The research involved 41 interviews with principal investigators, junior researchers, and technicians working at 13 different laboratories in the United States and Europe.

4. All names in this section are pseudonyms.

5. SPLASH: http://www.splashcatalog.org/.

6. No data were collected as part of an institutional review board–approved study, but Meyer has written permission from the collaboration's lead investigator to discuss the workings of the project and his own role therein.

7. For a complete list, see http://www.genome.gov/Pages/About/OD/OPG/PublicationsUsingGAINGenotypingData-072211.pdf.

8. For a complete list, see http://www.genome.gov/Pages/About/OD/OPG/GAIN Bibliography-021411.pdf.

9. DSM is the *Diagnostic and Statistical Manual of Mental Disorders*, which lists different categories of mental disorders and gives specific conditions required for a set of symptoms to "meet criteria." These criteria generally include a list of potential symptoms and number of symptoms required, plus the number of days the episode must have lasted to meet criteria. The DSM-IIIR was published in 1987 as a revised version of the 1980 DSM-III, and DSM-IV was published in 1994. The RDC (Research Diagnostic Criteria) is a similar, older system developed in the 1970s.

10. So, too, of course, do private companies, but this is a separate issue; see Savage and Burrows (2007, 2009). Portions of this section are based on Axelsson and Schroeder (2009).

11. See, for example, http://www.vr.se/download/18.227c330c123c73dc586800013477/1340207480682/NordicConferenceRInov2008.pdf.

12. Official Statistics of Sweden website: http://www.scb.se/en_/.

13. To gain insights into the possibilities and constraints of increased data sharing in Sweden, a number of interviews (16) were conducted with database managers, researchers using databases, and officials at funding bodies dealing with databases, specifically in relation to e-research and data sharing. The interviews in this section were carried out during the period of May 2006 to May 2008 in Umeå, Stockholm, and Gothenburg by Ralph Schroeder and Anne-Sofie Axelsson.

14. CENS: http://research.cens.ucla.edu/.

Chapter 6

1. An earlier version of this discussion was included in Meyer, Oostveen, Schroeder, et al. (2012). Copyright © Members of the SESERV Consortium and included with permission.

2. EGEE: https://eu-egee-org.web.cern.ch/.

3. Portions of this section are based on Schroeder (2008).

4. EGI: http://www.egi.eu/

5. EGEE built on several previous European efforts, including GÉANT, the European academic computing network, the European DataGrid project and the LHC Computing Grid (see http://lcg.web.cern.ch/LCG/), which shared computing resources before 2004. The organization of large-scale collaborative research organization in physics is analyzed in Shrum, Genuth, and Chompalov (2007).

6. For a discussion of GridPP, the UK particle physics e-science Grid, see Zheng, Venters, and Cornford (2011).

7. Portions of this section are based on Schroeder (2008).

8. IVOA: http://www.ivoa.net/.

9. As of January 2013.

10. IVOA Documents: http://www.ivoa.net/Documents/latest/IVOAParticipation.html. Note that open access to data is subject to being "commensurate with national or project-imposed proprietary periods."

11. One feature that makes the integration of IVOA data easier than other types of data is that it has no inherent commercial value.

12. See, for example, Mixed Media Grid: http://mimeg.sourceforge.net; and Understanding New Forms of Digital Record for e-Social Science: http://www.nottingham.ac.uk/cral/projects/dressi.aspx.

13. NCeSS: formerly at http://www.ncess.ac.uk/, but the site was taken down in January 2013.

14. In 2009, the NCeSS program was replaced by the national Office of the Director of Digital Social Research, a post held by Professor David De Roure. See http://www.digitalsocialresearch.net for current information.

15. DGrid: http://www.d-grid.de/.

16. Portions of this section were originally included in Schroeder (2008).

17. VOSON: http://voson.anu.edu.au/; discussed briefly in chapter 1.

18. NodeXL: http://nodexl.codeplex.com/.

19. Social Media Research Foundation: http://www.smrfoundation.org/.

20. Another possibility is that environmental activists have only an online presence, are organized only via their online links, and are focused on pushing a particular agenda online. But this possibility merely pushes the problem one step further: What is the importance of these relations in the websphere for the offline world generally?

21. *Cybermetrics*: http://cybermetrics.cindoc.csic.es/.

22. Portions of this section are based on Schroeder and Meyer (2012).

Chapter 7

1. Portions of this section are based on Meyer (2011).

2. See, for instance, William Pannapacker's (2009, 2011) descriptions of the growing energy around digital humanities at the Modern Language Association annual meetings.

3. Digital Image Archive of Medieval Music: http://www.diamm.ac.uk/.

4. Portions of this section are based on Meyer (2011).

5. Siobhan Davies RePlay (http://www.siobhandaviesreplay.com) is a digital archive of dance resources. Siobhan Davies is a leading UK choreographer, and the RePlay archive contains more than 5,000 items relating to her choreography, dancing, and dance training.

6. OBPO: http://www.oldbaileyonline.org/. OBPO contains all published accounts of trials held at the Old Bailey court (the Central Criminal Court of England and Wales) from 1674 to 1913 ($n = 197{,}745$). The content includes 127 million words of tagged and accurately transcribed historical text, the largest "fully searchable edition ... detailing the lives of non-elite people" ever published (from the OBPO website, http://www.oldbaileyonline.org/, accessed October 28, 2010).

7. Oxford University podcast collection: http://podcasts.ox.ac.uk/ and http://itunes.ox.ac.uk. The Oxford podcasts are distributed through multiple outlets (reflecting the university's decentralized nature), but in 2008, spurred by a partnership with Apple's iTunes U podcasting platform, the university's computing service began to more systematically provide support for distribution of podcasts via iTunes U as well as via University of Oxford Podcasts (http://podcasts.ox.ac.uk), the University of Oxford website (http://itunes.ox.ac.uk/), and the Mobile Oxford portal (http://m.ox.ac.uk/). The complete collection at the time of this project consisted of approximately 2,650 podcasts, including lectures, interviews, discussions, and workshops. Two-thirds of the files are audio, and the remaining are primarily video plus a few electronic books and portable documents.

8. The Online Historical Population Reports project, which goes by the short name "Histpop" (http://www.histpop.org), was publicly launched in 2007. It contains some 200,000 pages of census and registration material for the British Isles and provides free online access to the complete British population reports for Britain and Ireland from 1801 to 1937.

9. The website A Vision of Britain through Time (http://www.visionofbritain.org.uk/) was originally launched in 2004 using National Lottery funding. The site includes geographical surveys of Britain, including every census from 1801 to 2001, and a library of historic maps. The site was designed mainly with local historians in mind as the key user base.

10. However, a lack of local support can easily negate these gains, such as for the survey respondent who indicated that although "all of my students should listen to [the interview with Davies] ... I am not sufficiently adept to put it on a computer and project it, partly because it is so awkward to do so in this particular college" (qtd. in Meyer 2011, 35).

11. Portions of this section are based on Schroeder and Meyer (2012).

12. See http://www.w3.org/Provider/Style/URI for a description of the logic and theory behind the Cool URI.

13. For Goodman's talk, see http://www.oii.ox.ac.uk/events/?id=571.

14. ADS Labs: http://labs.adsabs.harvard.edu/.

15. Worldwide Telescope: http://www.worldwidetelescope.org/.

16. Portions of this section are based on Meyer and Schroeder (2009b).

Chapter 8

1. Portions of this section are based on Schroeder (2007a).

2. See http://www.omii.ac.uk/ for the Open Middleware Infrastructure Institute's mission statement: "OMII-UK is an open-source organisation that empowers the U.K. research community by providing software for use in all disciplines of research. Our mission is to cultivate and sustain community software important to research. All of OMII-UK's software is free, open source and fully supported."

3. OSS Watch: http://www.oss-watch.ac.uk/.

4. Science Commons: http://sciencecommons.org/.

5. In the sense of a common pool of resources, although not entirely because Creative Commons licenses are not subtractable in the same sense that fisheries or grazing areas are, for instance.

6. Research Data Alliance: http://rd-alliance.org/.

7. Merton's other two norms, disinterestedness and organized skepticism, are also relevant, and all four are interrelated, but these two are perhaps less directly relevant to e-research.

8. For the distinctions between different types of licenses, see Weber (2004).

9. One lesson from open source for open e-research is that it is necessary to protect research and tools in order to release them for free access and use. For open source, this protection is afforded by means of licensing, either to prevent commercial use or to ensure that even commercial use stays "open." For academic research, one way to safeguard ownership or credit is by means of copyrighting and developing various forms of licensing modeled on open source as a means of disseminating research (not only publications, but also data and tools). This key element ensures the "open science" character of e-research (Willinsky 2005b; Burk 2007).

10. See Willinsky (2005a) for an inventory of open-access efforts current at the time.

11. The umbrella term *intellectual property* conflates several phenomena that are separate from a legal perspective: copyright, patents, and database protection (two additional areas of intellectual property, trademarks and trade secrets, are less relevant in this context). There is no space to analyze the implications of these differences, especially as the issues involved, especially in the digital realm, are both

complex and in flux. In brief, however, the implications of these distinctions will be different if one takes either a legal and economic perspective (e.g., asking to what extent the protection or use of ideas fosters or hinders economic innovation) or, as here, a user perspective (e.g., asking to what extent content and tools are free of charge, accessible, usable, modifiable, and transportable).

12. Martin Bauer and George Gaskell's functionalist model is in fact more complicated because it deals with an area (biotechnology) where there has been considerable controversy and thus also involves the media and commercial interests. e-Research is more contained within the research community and among NGOs and advocacy groups.

13. For the state's role in the globalization of science, see Drori, Meyere, Ramirez, et al. (2003). The global nature of e-research is also discussed in Schroeder (2006). On the state's continuing strength and its role as a target of social movements, see Mann (2013).

14. The literature on these phenomena is extensive, but Van der Vleuten (2004) provides a good overview in relation to large technological systems and infrastructures.

15. Named after the biblical book of Matthew, in particular the passage from chapter 25, verse 29: "For unto every one that hath shall be given, and he shall have abundance: but from him that hath not shall be taken away even that which he hath" (qtd. in Merton 1968, 58).

16. This user base points to an important distinction between a bottom-up approach to infrastructures by means of sharing computing resources (as with the Seti@home research project [http://setiathome.berkeley.edu/], whereby the spare computing capacity of Internet-connected PCs was harnessed for the Search for Extraterrestrial Intelligence) or by means of user-generated content (for example, *Wikipedia* [see Benkler 2006]) in a decentralized way and top-down approaches, which are characteristic, for example, of national communication technology infrastructures (or electricity grids). Note, however, that both require more or less centralized coordination and control rather than being at one or other extreme.

17. For the different social science perspectives on e-science, see Schroeder and Fry (2007).

18. In this respect, perhaps the most critical controversy is the current challenge to foster "net neutrality," which may mean the difference between an e-infrastructure that will come to be multitiered rather than being part of a single network (Benkler 2006, 396–397): whether the Internet will continue to operate on the "end-to-end" principle whereby content is delivered regardless of its origin and destination or if different parts come to operate according to different principles.

19. Part of the reason that the arguments creating these tensions are not simply for or against open e-infrastructures is that they operate on different levels or in differ-

ent realms: the realm of legal regulation, of the technical standard-setting possibilities for open systems, and of research-funding bodies and advocacy movements, for example, will push and pull the system on these different levels—although the point of large technological systems, of course, is that they will ultimately have to become aligned or gel into a stable whole.

20. This tension has been noticed for basic research: Why should one country support basic research only to see another country exploit this research?

21. This is akin to the "shadow of the future" in open-source software: potential commercial reward in the longer term if there are many users. See Weber (2004).

22. Though whether scientific e-research can be independent of political borders for countries such as China, which is developing an e-research infrastructure and yet imposes restricted Internet and Web access, remains to be seen.

Chapter 9

1. In the second phase of funding, the project was renamed GENeSIS and included a new partnership with the University of Leeds.

2. Portions of this section are based on Meyer, Schroeder, and Dutton (2008).

3. Copyright, Designs and Patents Act (Great Britain) (1988), c. 48 § X, p. 163.

4. Copyright Law of the United States of America (1976), 17 U.S.C. § 105.

5. For the *Guardian*'s campaign, see http://www.guardian.co.uk/technology/freeourdata.

6. Portions of this section are based on Meyer, Thomas, and Schroeder (2011).

7. Funding for our research on the topic of web archives includes the JISC/National Endowment for the Humanities Transatlantic Digitisation Collaboration: World Wide Web of Humanities (PI Eric T. Meyer, 2008–2009, http://www.oii.ox.ac.uk/research/?id=48), JISC Researcher Engagement with Web Archives (PI Eric T. Meyer, 2010, http://www.oii.ox.ac.uk/research/?id=60), International Internet Preservation Consortium Using Web Archives: A Futures Perspective (PI Eric T. Meyer, 2011, http://ssrn.com/abstract=1830025), JISC Big Data: Demonstrating the Value of the UK Web Domain Dataset for Social Science Research (PI Helen Margetts, 2012–2013, http://www.oii.ox.ac.uk/research/projects/?id=88), and Arts and Humanities Research Council Big UK Domain Data for the Arts and Humanities (PI Jane Winters, 2014–2015, http://buddah.projects.history.ac.uk/).

8. JISC (now Jisc, http://www.jisc.ac.uk/) funded information and communication technologies research and infrastructure development in the education and research sector in the United Kingdom.

9. Of course, this is potentially happening at exactly the time when actual dusty archives are being digitized and made available on the Web, creating the possibility that many born-printed materials might become more widely available online than certain born-digital materials.

10. Internet Archive's Wayback Machine: http://archive.org/web/web.php.

11. For instance, the Sunbelt conference, which is a leading conference for scholars interested in SNA, held its thirty-third annual meeting in 2013.

Chapter 10

1. Portions of this section are based on Meyer and Schroeder (2013).

2. None of the cases directly represent style 1, postulation, but in some sense they all do because this style is related to the mathematization of disciplines we discuss elsewhere. Style 6, historicogenetic, is also not directly represented in this selection of cases. As mentioned in chapter 2, these two styles are less closely linked with e-research than the other styles are.

3. It can be noted that we use the term *advancing* here in the sense of moving forward; no normative implications are being suggested. On this point, see (Schroeder 2007b).

4. Kling, McKim, and King (2003, 58) argue that the term *architectural choice point* "refers to a technological feature or social arrangement in which the designer can select alternatives." This part of a STIN consists of inflection points at which changes to a system are most likely to be introduced.

5. Portions of this section are based on Meyer, Schroeder, and Dutton (2008).

6. Ocean Tracking Network: http://oceantrackingnetwork.org/.

7. One example of the inadequacy of current ideas in science and technology studies in this regard can be given. In his recent book about climate models, Paul Edwards (2010) mentions the volunteer computing tool climateprediction.net (http://www.climateprediction.net/), which falls into our definition of e-research. But he points out that current theories of science and technology studies, unlike the "realist" account of science we give here, do not allow for the validity or otherwise of climate knowledge, arguing instead that scientific knowledge is socially constructed, that "social agreement is sufficient for knowledge production" (437). Put differently, there are no independent criteria for the validity of knowledge. But this assessment clearly makes Edwards, a science and technology studies scholar, uneasy, so his conclusion wobbles on this point: "we have few good reasons to doubt these facts" about climate change "and many reasons to trust their validity" (439): a realist can rely on these facts or data here even if the facts must also continue to be scrutinized and can trust scientific knowledge exclusively rather than the wobbly "many reasons."

References

Aarseth, Espen J. 1997. *Cybertext: Perspectives on Ergodic Literature*. Baltimore: Johns Hopkins University Press.

Ackland, Robert. 2009. Social Network Services as Data Sources and Platforms for e-Researching Social Networks. *Social Science Computer Review* 27 (4): 481–492.

Ackland, Robert, Mathieu O'Neil, Bruce Bimber, Rachel Gibson, and Stephen Ward. 2006. New Methods for Studying Online Environmental-Activist Networks. Paper presented at the Twenty-Sixth International Sunbelt Social Network Conference, Vancouver, Canada, April 24–30.

Ackland, Robert, Mathieu O'Neil, Russell Standish, and Markus Buchhorn. 2006. VOSON: A Web Services Approach for Facilitating Research into Online Networks. Paper presented at the Second International Conference on e–Social Science, University of Manchester, Manchester, UK, June 28–30.

Altintas, Ilkay, Chad Berkley, Efrat Jaeger, Matthew Jones, Bertram Ludascher, and Steve Mock. 2004. Kepler: An Extensible System for Design and Execution of Scientific Workflows. In *Proceedings of the 16th International Conference on Scientific and Statistical Database Management (SSDBM '04)*, 423–424. New York: IEEE.

Anderson, Chris. 2006. *The Long Tail: How Endless Choice Is Creating Unlimited Demand*. London: Random House Business Books.

Atkins, Daniel E., Kelvin K. Droegemeier, Stuart I. Feldman, Hector Garcia-Molina, Michael L. Klein, David G. Messerschmitt, Paul Messina, Jeremiah P. Ostriker, and Margaret H. Wright. 2003. *Revolutionizing Science and Engineering through Cyberinfrastructure: Report of the National Science Foundation Blue-Ribbon Advisory Panel on Cyberinfrastructure*. Washington, DC: National Science Foundation.

Aucott, Paul, Richard Healey, and Humphrey Southall. 2011. *Case Study: Embedding a Vision of Britain through Time as a Resource for Academic Research and Learning*. London: JISC.

Axelsson, Anne-Sofie, and Ralph Schroeder. 2009. Making It Open and Keeping It Safe: e-Enabled Data Sharing in Sweden. *Acta Sociologica* 52 (3): 213–226.

Baehr, Steven, Arun Vedachalam, Kirk D. Borne, and Daniel Sponseller. 2010. Data Mining the Galaxy Zoo Mergers. In *Proceedings of the 2010 Conference on Intelligent Data Understanding (CIDU 2010)*, ed. Ashok N. Srivastava, Nitesh Chawla, Philip S. Yu, and Paul Melby, 133–144. Mountain View, California: NASA.

Bairoch, A. 2000. Serendipity in Bioinformatics: The Tribulations of a Swiss Bioinformatician through Exciting Times! *Bioinformatics (Oxford, England)* 16 (1): 48–64.

Barjak, Franz, Kathryn Eccles, Eric T. Meyer, Simon Robinson, and Ralph Schroeder. 2013. The Emerging Governance of e-Infrastructure. *Journal of Computer-Mediated Communication* 18 (2): 113–136.

Barjak, Franz, Xuemei Li, and Mike Thelwall. 2007. Which Factors Explain the Web Impact of Scientists' Personal Homepages? *Journal of the American Society for Information Science and Technology* 58 (2): 200–211.

Barjak, Franz, Gordon Wiegand, Julia Lane, Zack Kertcher, Rob Procter, Meik Poschen, Simon Robinson, and Alexander Mentrup. 2008. *Accelerating Transition to Virtual Research Organisation in Social Science (AVROSS): Final Report*. Brussels: Information Society and Media Directorate General, Commission of the European Communities.

Bauer, Martin W., and George Gaskell. 2002. The Biotechnology Movement. In *Biotechnology: The Making of a Global Controversy*, ed. Martin W. Bauer and George Gaskell, 379–404. Cambridge, UK: Cambridge University Press.

Becher, Tony, and Paul Trowler. 2001. *Academic Tribes and Territories: Intellectual Enquiry and the Cultures of Disciplines*. 2nd ed. Milton Keynes, UK: Society for Research into Higher Education & Open University Press.

Benkler, Yochai. 2006. *The Wealth of Networks*. New Haven, CT: Yale University Press.

Berman, Francine D., and Henry E. Brady. 2005. *Final Report NSF SBE–CISE Workshop on Cyberinfrastructure and the Social Sciences*. Washington, DC: National Science Foundation.

Berman, Francine D., Geoffrey Fox, and Tony Hey. 2003. *Grid Computing: Making the Global Infrastructure a Reality*. Hoboken, NJ: Wiley.

Bijker, Wiebe E. 1995. *Of Bicycles, Bakelites, and Bulbs: Toward a Theory of Sociotechnical Change*. Cambridge, MA: MIT Press.

Bijker, Wiebe E. 2001. Social Construction of Technology. In *International Encyclopedia of the Social & Behavioral Sciences*, ed. Neil J. Smelser and Paul B. Baltes, 15522–15527. Oxford: Elsevier Science.

Bijker, Wiebe E., Thomas P. Hughes, and Trevor Pinch. 1987. *The Social Construction of Technological Systems: New Directions in the Sociology and History of Technology*. Cambridge, MA: MIT Press.

Blaney, Jonathan, and Peter Webster. 2010. *Rapid Impact Analysis: The Impact and Embedding of an Established Resource: British History Online as a Case Study.* London: JISC.

Boczkowski, Pablo J., and Leah A. Lievrouw. 2008. Bridging STS and Communication Studies: Scholarship on Media and Information Technologies. In *The Handbook of Science and Technology Studies,* ed. Edward J. Hackett, Olga Amsterdamska, Michael Lynch, and Judy Wajcman, 949–977. Cambridge, MA: MIT Press.

Borgman, Christine L. 2005. Disciplinary Differences in e-Research: An Information Perspective. Keynote talk at the First International Conference on e-Social Science, Manchester, UK, June 22–24.

Borgman, Christine L. 2007. *Scholarship in the Digital Age: Information, Infrastructure, and the Internet.* Cambridge, MA: MIT Press.

Borgman, Christine. 2015. *Big Data, Little Data, No Data: Scholarship in the Networked World.* Cambridge, MA: MIT Press.

Borgman, Christine L., Jillian C. Wallis, Matthew S. Mayernik, and Alberto Pepe. 2007. Drowning in Data: Digital Library Architecture to Support Scientific Use of Embedded Sensor Networks. In *Proceedings of the 7th Joint Conference on Digital Libraries,* ed. Edie Rasmussen, Ray R. Larson, Elaine Toms, and Shigeo Sugimoto, 269–277. New York: Association for Computing Machinery.

Börner, Katy, Chaomei Chen, and Kevin W. Boyack. 2005. Visualizing Knowledge Domains. *Annual Review of Information Science & Technology* 37 (1): 179–255.

Bos, Nathan, Ann Zimmerman, Judith Olson, Jude Yew, Jason Yerkie, Erik Dahl, and Gary Olson. 2007. From Shared Databases to Communities of Practice: A Taxonomy of Collaboratories. *Journal of Computer-Mediated Communication* 12 (2):652–672.

Brabazon, Tara. 2006. The Google Effect: Googling, Blogging, Wikis, and the Flattening of Expertise. *Libri* 56:157–167.

Brower, Ralph S. 2000. On Improving Qualitative Methods in Public Administration Research. *Administration & Society* 32 (4): 363–397.

Bryant, Susan L., Andrea Forte, and Amy Bruckman. 2005. Becoming Wikipedian: Transformation of Participation in a Collaborative Online Encyclopedia. In *Proceedings of the 2005 International ACM SIGGROUP conference on Supporting Group Work,* ed. Kjeld Schmidt, Mark Pendergast, Mark Ackerman, and Gloria Mark, 1–10. New York: Association for Computing Machinery.

Bulger, Monica, Eric T. Meyer, Grace de la Flor, Melissa Terras, Sally Wyatt, Marina Jirotka, Kathryn Eccles, and Christine Madsen. 2011. *Reinventing Research? Information Practices in the Humanities.* London: Research Information Network.

Burk, Dan L. 2007. Intellectual Property in the Context of e-Science. *Journal of Computer-Mediated Communication* 12 (2): 600–617.

Bush, Vannevar. 1945. Science: The Endless Frontier. *Transactions of the Kansas Academy of Science* 48 (3): 231–264.

Calambokidis, John. 2010. *Symposium on the Results of the SPLASH Humpback Whale Study: Final Report and Recommendations*. Olympia, WA: Cascadia Research.

Calambokidis, John, Jay Barlow, Alexander M. Burdin, Phillip Clapham, John K. B. Ford, Christine M. Gabriele, David Mattila, et al. 2007. New Insights on Migrations and Movements of North Pacific Humpback Whales from the SPLASH Project. Paper presented at the Biennial Conference on the Biology of Marine Mammals, Cape Town, South Africa, November 29–December 3.

Calambokidis, John, Erin A. Falcone, Terrance J. Quinn, Alexander M. Burdin, Phillip Clapham, John K. B. Ford, Christine M. Gabriele, et al. 2008. *SPLASH: Structure of Populations, Levels of Abundance, and Status of Humpback Whales in the North Pacific: Final Report for Contract AB133F–03-RP-00078*. Seattle, WA: US Department of Commerce.

Caldas, Alexandre, Ralph Schroeder, G. Mesch, and William H. Dutton. 2008. Patterns of Information Search and Access on the World Wide Web: Democratizing Expertise of Creating New Hierarchies? *Journal of Computer-Mediated Communication* 13 (4): 769–793.

Callon, Michel. 1987. Society in the Making: The Study of Technology as a Tool for Social Analysis. In *The Social Construction of Technological Systems*, ed. Wiebe E. Bijker, Thomas P. Hughes, and Trevor Pinch, 83–103. Cambridge, MA: MIT Press.

Cardamone, Carolin, Kevin Schawinski, Marc Sarzi, Steven P. Bamford, Nicola Bennert, C. M. Urry, Chris Lintott, et al. 2009. Galaxy Zoo Green Peas: Discovery of a Class of Compact Extremely Star-Forming Galaxies. *Monthly Notices of the Royal Astronomical Society* 399 (3): 1191–1205.

Carlson, Samuelle, and Ben Anderson. 2006. e-Nabling Data: Potential Impacts on Data, Methods, and Expertise. Paper presented at the Second International Conference on e-Social Science, Manchester, UK, June 28–30.

Carlson, Samuelle, and Ben Anderson. 2007. What *Are* Data? The Many Kinds of Data and Their Implications for Data Re-use. *Journal of Computer-Mediated Communication* 12 (2): 635–651.

Caro, Robert. 2012. *The Passage of Power: The Years of Lyndon Johnson*. Vol. 4. London: Bodley Head.

Carusi, Annamaria, and Marina Jirotka. 2009. From Data Archive to Ethical Labrynth. *Qualitative Research* 9 (3): 285–298.

Carusi, Annamaria, Marina Jirotka, and Michael Parker. 2006. *Trust and Ethics in e-Science Agenda Setting: Workshop Report*. Oxford: Oxford e-Research Centre.

Centre for Information Behaviour and the Evaluation of Research. 2008. *Information Behaviour of the Researcher of the Future*. CIBER Briefing Paper. London: University College London.

Ceruzzi, Paul. 1999. *A History of Modern Computing*. Cambridge, MA: MIT Press.

Collins, Ellen, Monica Bulger, and Eric T. Meyer. 2011. Discipline Matters: Technology Use in the Humanities. *Arts and Humanities in Higher Education* 11 (1–2): 76–92.

Collins, Randall. 1993. Ethical Controversies of Science and Society: A Relation between Two Spheres of Social Conflict. In *Controversial Science: From Content to Contention*, ed. Thomas Brante, Steve Fuller, and William Lynch, 301–317. Albany: State University of New York Press.

Collins, Randall. 1994. Why the Social Sciences Won't Become High-Consensus, Rapid-Discovery Science. *Sociological Forum* 9 (2): 155–177.

Collins, Randall. 1998. *The Sociology of Philosophies: A Global Theory of Intellectual Change*. Cambridge, MA: Belknap Press of Harvard University Press.

Consalvo, Mia, and Charles Ess, eds. 2011. *The Handbook of Internet Studies*. Malden, MA: Wiley-Blackwell.

Cooper, Geoff, and John Bowers. 1995. Representing the User: Notes on the Disciplinary Rhetoric of Human–Computer Interaction. In *The Social and Interactional Dimensions of Human–Computer Interfaces*, ed. Peter J. Thomas, 48–66. Cambridge, UK: Cambridge University Press.

Crombie, A. C. 1994. *Styles of Scientific Thinking in the European Tradition: The History of Argument and Explanation Especially in the Mathematical and Biomedical Sciences and Arts*. London: Duckworth.

Crowe, Elizabeth Powell. 2008. *Genealogy Online*. 8th ed. Columbus, OH: McGraw-Hill.

Cummings, Jonathon N., and Sara Kiesler. 2005. Collaborative Research across Disciplinary and Organizational Boundaries. *Social Studies of Science* 35 (5): 703–722.

Cunningham, Ward. 2012. Wiki History. Last modified October 8. http://c2.com/cgi/wiki?WikiHistory. Accessed January 15, 2013.

Dagens Nyheter. 1986. 15,000 svenskar i hemligt dataregister (15,000 Swedes in secret computer register). February 10.

David, Paul A. 2004. Understanding the Emergence of "Open Science" Institutions: Functionalist Economics in Historical Context. *Industrial and Corporate Change* 13 (4): 571–589.

David, Paul A., Matthijs den Besten, and Ralph Schroeder. 2010. Open Science and e-Science. In *World Wide Research: Reshaping the Sciences and Humanities*, ed. William H. Dutton and Paul Jeffreys, 299–316. Cambridge, MA: MIT Press.

David, Paul A., and Bronwyn H. Hall. 2006. Property and the Pursuit of Knowledge: IPR Issues Affecting Scientific Research. *Research Policy* 35 (6): 767–771.

David, Paul A., and Michael Spence. 2003. *Towards Institutional Infrastructures for e-Science: The Scope of the Challenge*. Oxford Internet Institute Research Report no. 2. Oxford, UK: Oxford Internet Institute.

David, Paul A., and Paul F. Uhlir. 2005. Creating the Information Commons for e-Science: Toward Institutional Policies and Guidelines for Action. *CODATA Newsletter* 91 (July): 1–3. http://www.codata.org/resources/newsletters/newsltr91A4.pdf. Accessed August 8, 2012.

Dellavalle, Robert P., Eric J. Hester, Lauren F. Heilig, Amanda L. Drake, Jeff W. Kuntzman, Marla Graber, and Lisa M. Schilling. 2003. Going, Going, Gone: Lost Internet References. *Science* 302 (5646): 787–788.

Demeritt, David. 2000. The New Social Contract for Science: Accountability, Relevance, and Value in US and UK Science and Research Policy. *Antipode* 32 (3): 308–329.

Demographic Data Base. 2009. Website. http://www.ddb.umu.se/english/?languageId=1. Accessed March 19, 2009.

Den Besten, Matthijs, Arthur Thomas, and Ralph Schroeder. 2009. Life Science Research and Drug Discovery at the Turn of the 21st Century: The Experience of SwissBioGrid. *Journal of Biomedical Discovery and Collaboration* 4: article 5.

De Solla Price, Derek J. 1963. *Little Science, Big Science*. New York: Columbia University Press.

Dillman, Don. 2000. *Mail and Internet Surveys: The Tailored Design Method*. 2nd ed. New York: Wiley.

Dougherty, Meghan, Eric T. Meyer, Christine Madsen, Charles Van den Heuvel, Arthur Thomas, and Sally Wyatt. 2010. *Researcher Engagement with Web Archives: State of the Art*. London: JISC.

Drori, Gil S., John W. Meyer, Francisco O. Ramirez, and Evan Schofer. 2003. *Science in the Modern World Polity: Institutionalization and Globalization*. Stanford, CA: Stanford University Press.

Duguid, Paul. 2007. Inheritance and Loss? A Brief Survey of Google Books. *First Monday* 12 (8).

Duhigg, Charles. 2012a. How Companies Learn Your Secrets. *New York Times*, February 16.

Duhigg, Charles. 2012b. *The Power of Habit: Why We Do What We Do in Life and Business*. New York: Random House.

Dutton, William H. 2005. The Internet and Social Transformation: Reconfiguring Access. In *Transforming Enterprise: The Economic and Social Implications of Information Technology*, ed. William H. Dutton, Brian Kahin, Ramon O'Callaghan and Andrew W. Wyckoff, 375–397. Cambridge, MA: MIT Press.

Dutton, William H. 2011. The Politics of Next Generation Research: Democratizing Research-Centred Computational Networks. *Journal of Information Technology* 26: 109–119.

Dutton, William H., ed. 2013. *The Oxford Handbook of Internet Studies*. Oxford: Oxford University Press.

Dutton, William H., and Grant Blank. 2011. *Next Generation Users: The Internet in Britain*. Oxford Internet Survey 2011. Oxford, UK: Oxford Internet Institute, University of Oxford.

Dutton, William H., Annamaria Carusi, and Malcolm Peltu. 2006. Fostering Multidisciplinary Engagement: Communication Challenges for Social Research on Emerging Digital Technologies. *Prometheus* 24 (2): 129–149.

Dutton, William H., and Paul Jeffreys. 2010a. World Wide Research: An Introduction. In *World Wide Research: Reshaping the Sciences and Humanities*, ed. William H. Dutton and Paul Jeffreys, 1–17. Cambridge, MA: MIT Press.

Dutton, William H., and Paul Jeffreys, eds. 2010b. *World Wide Research: Reshaping the Sciences and Humanities*. Cambridge, MA: MIT Press.

Dutton, William H., and Eric T. Meyer. 2009. Experience with New Tools and Infrastructures of Research: An Exploratory Study of Distance from, and Attitudes toward, e-Research. *Prometheus* 27 (3): 223–238.

Dutton, William H., and A. Shepherd. 2006. Trust in the Internet as an Experience Technology. *Information Communication and Society* 9 (4): 433–451.

Dyson, Esther. 1997. *Release 2.0: A Design for Living in the Digital Age*. New York: Broadway Books.

Eden, Grace, Marina Jirotka, and Eric T. Meyer. 2012. Interpreting Digital Images beyond Just the Visual: Crossmodal Practices in Medieval Musicology. *Interdisciplinary Science Reviews* 37 (1): 69–85.

Edwards, Paul N. 2010. *A Vast Machine: Computer Models, Climate Data, and the Politics of Global Warming*. Cambridge, MA: MIT Press.

Evans, James, and Andrey Rzhetsky. 2010. Machine Science. *Science* 329 (5990): 399–400.

Eysenbach, Gunther, John Powell, Oliver Kuss, and Eun-Ryoung Sa. 2002. Empirical Studies Assessing the Quality of Health Information for Consumers on the World Wide Web: A Systematic Review. *Journal of the American Medical Association* 287 (20): 2691–2700.

Fenn, Jackie, and Mark Raskino. 2008. *Mastering the Hype Cycle*. Cambridge, MA: Harvard Business School Press.

Fielding, Nigel, Raymond M. Lee, and Grant Blank, eds. 2008. *The Sage Handbook of Online Research Methods*. London: Sage.

Finholt, Thomas A. 2003. Collaboratories as a New Form of Scientific Organization. *Economics of Innovation and New Technology* 12 (1): 5–25.

Fischer, Michael, Stephen Lyon, and David Zeitlyn. 2008. The Internet and the Future of Social Science Research. In *The Sage Handbook of Online Research Methods*, ed. Nigel Fielding, Raymond M. Lee, and Grant Blank, 519–536. Thousand Oaks, CA: Sage.

Fish, Stanley. 2012. Mind Your P's and B's: The Digital Humanities and Interpretation. Last modified January 23. http://opinionator.blogs.nytimes.com/2012/01/23/mind-your-ps-and-bs-the-digital-humanities-and-interpretation/. Accessed January 24, 2012.

Fore, Joe, Ilse R. Wiechers, and Robert Cook-Deegan. 2006. The Effects of Business Practices, Licensing, and Intellectual Property on Development and Dissemination of the Polymerase Chain Reaction: Case Study. *Journal of Biomedical Discovery and Collaboration* 1: article 7.

Foster, Ian, and Carl Kesselman, eds. 2004. *The Grid: Blueprint for a New Computing Infrastructure*. 2nd ed. San Francisco: Elsevier and Morgan Kaufmann.

Freeman, Linton C. 2004. *The Development of Social Network Analysis: A Study in the Sociology of Science*. Vancouver: Empirical Press.

Freeman, Linton C. 2008. Going the Wrong Way on a One-Way Street: Centrality in Physics and Biology. *Journal of Social Structure* 9: (2).

Free Our Data (blog). 2011. Environment Agency ... Sells Our Data? Last modified June 15. http://www.freeourdata.org.uk/blog/2011/06/environment-agency-sells-our-data/. Accessed January 15, 2012.

Frickel, Scott, and Neil Gross. 2005. A General Theory of Scientific/Intellectual Movements. *American Sociological Review* 70 (2): 204–232.

Fry, Jenny. 2006. Scholarly Research and Information Practices: A Domain Analytic Approach. *Information Processing & Management* 42 (1): 299–316.

Fry, Jenny, and Ralph Schroeder. 2010. Disciplinary Differences in e-Research. In *World Wide Research: Reshaping the Sciences and Humanities*, ed. William H. Dutton and Paul Jeffreys, 257–275. Cambridge, MA: MIT Press.

References

Fry, Jenny, Ralph Schroeder, and Matthijs den Besten. 2009. Open Science in e-Science: Contingency or Policy? *Journal of Documentation* 65 (1): 6–32.

Fry, Jenny, and Sanna Talja. 2004. The Cultural Shaping of Scholarly Communication: Explaining e-Journal Use within and across Academic Fields. *Proceedings of the American Society for Information Science and Technology* 41 (1): 20–30.

Fry, Jenny, Shefali Virkar, and Ralph Schroeder. 2008. Search Engines and Expertise about Global Issues: Well-Defined Territory or Undomesticated Wilderness? In *Websearch: Interdisciplinary Perspectives*, edited by Michael Zimmer and Amanda Spink, 255–276. Berlin: Springer.

Fuchs, Stephan. 2001a. *Against Essentialism: A Theory of Culture and Society*. Cambridge, MA: Harvard University Press.

Fuchs, Stephan. 2001b. What Makes Sciences "Scientific"? In *Handbook of Sociological Theory*, edited by Jonathan H. Turner, 21–35. New York: Springer Science+Business Media.

Gagliardi, Fabrizio. 2005. The EGEE European Grid Infrastructure Project. In *High Performance Computing for Computational Science—VECPAR 2004 LNCS 3402*, ed. Michel Dayde, Jack Dongarra, Vincente Hernandez, and Jose M. L. M. Palma, 194–203. Berlin: Springer.

Galison, Peter. 1997. *Image and Logic: A Material Culture of Microphysics*. Chicago: University of Chicago Press.

Galison, Peter, and Bruce William Hevly, eds. 1992. *Big Science: The Growth of Large-Scale Research*. Stanford, CA: Stanford University Press.

Gardner, Daniel, Arthur W. Toga, Giorgio A. Ascoli, Jackson T. Beatty, James F. Brinkley, Anders M. Dale, Peter T. Fox, et al. 2003. Towards Effective and Rewarding Data Sharing. *Neuroinformatics* 1 (3): 289–295.

Gardner, David. 2007. Whale Survives Harpoon Attack 130 Years Ago to Become "World's Oldest Mammal." *Daily Mail Online (U.K.)*, June 13.

Garvey, William D., and Belver C. Griffith. 1967. Scientific Communication as a Social System. *Science* 157 (3792): 1011–1016.

Garvey, William D., and Belver C. Griffith. 1972. Communication and Information Processing within Scientific Disciplines: Empirical Findings for Psychology. *Information Storage and Retrieval* 8 (3): 123–136.

Gifford, Don, and R. J. Seidman. 1974. *Notes for Joyce: An Annotation of James Joyce's Ulysses*. New York: Dutton.

Gläser, Jochen. 2003. What Internet Use Does and Does Not Change in Scientific Communities. *Science Studies* 16 (1): 38–51.

Gläser, Jochen. 2006. *Wissenschaftliche Produktionsgemeinschaften: Die Soziale Ordnung der Forschung*. Frankfurt: Campus.

Goertz, Gary, and James Mahoney. 2012. *A Tale of Two Cultures: Qualitative and Quantitative Research in the Social Sciences*. Princeton, NJ: Princeton University Press.

Goodman, Alyssa, and Chris Lintott. 2013. Seamless Astronomy, Sea Monsters, and the Milky Way. Paper presented at the Innovation and Digital Scholarship Lecture Series, Oxford Internet Institute, University of Oxford, February 21.

Gordon, Gerald, Sue Marquis, and O. W. Anderson. 1962. Freedom and Control in Four Types of Scientific Settings. *American Behavioral Scientist* 6 (4): 39–43.

Grant, J. Kerry. 2001. *A Companion to V*. Athens: University of Georgia Press.

Grant, J. Kerry. 2008. *A Companion to The Crying of Lot 49*. 2nd ed. Athens: University of Georgia Press.

Haas, Peter M. 1992. Introduction: Epistemic Communities and International Policy Coordination. *International Organization* 46 (1): 1–35.

Hackett, Edward J. 2005. Essential Tensions: Identity, Control, and Risk in Research. *Social Studies of Science* 35 (5): 787–826.

Hacking, Ian. 1983. *Representing and Intervening*. Cambridge: Cambridge University Press.

Hacking, Ian. 1992. The Self-Vindication of the Laboratory Sciences. In *Science as Practice and Culture*, ed. Andrew Pickering, 29–64. Chicago: University of Chicago Press.

Hacking, Ian. 2002. *Historical Ontology*. Cambridge, MA: Harvard University Press.

Hacking, Ian. 2009. *Scientific Reason*. Taipei: National Taiwan University Press.

Hacking, Ian. 2012. "Language, Truth and Reason" 30 Years Later. *Studies in History and Philosophy of Science* 43:599–609.

Hakken, David. 2003. *The Knowledge Landscapes of Cyberspace*. New York: Routledge.

Hale, Scott, Taha Yasseri, Josh Cowls, Eric T. Meyer, Ralph Schroeder, and Helen Margetts. 2014. Mapping the UK Webspace: Fifteen Years of British Universities on the Web. In *Proceedings of the 2014 ACM Conference on Web Science*, ed. Filippo Menczer, Jim Hendler, William Dutton, Markus Strohmaier, Eric T. Meyer, and Ciro Cattuto, 62–70. New York: Association for Computing Machinery.

Hallmark, Julie. 2004. Access and Retrieval of Recent Journal Articles: A Comparative Study of Chemists and Geoscientists. *Issues in Science and Technology Librarianship* 40: article 1.

Hara, Noriko, and Howard Rosenbaum. 2008. Revising the Conceptualization of Computerization Movements. *Information Society* 24 (4): 229–245.

Harman, Jay R. 2003. Whither Geography? *Professional Geographer* 55 (4): 415–421.

Harnad, Stevan. 2001. The Self-Archiving Initiative. *Nature* 410 (6832): 1024–1025.

Head, Alison J. 2007. Beyond Google: How Do Students Conduct Academic Research? *First Monday* 12 (8).

Heimeriks, Gaston, and Eleftheria Vasileiadou. 2008. Changes or Transition? Analysing the Use of ICTs in the Sciences. *Social Sciences Information / Information sur les sciences sociales* 47 (1): 5–29.

Heller, Agnes. 1989. From Hermeneutics in Social Science toward a Hermeneutics of Social Science. *Theory and Society* 18 (3): 291–322.

Helsper, Ellen J., and Rebecca Eynon. 2010. Digital Natives: Where Is the Evidence? *British Educational Research Journal* 36 (3): 503–520.

Henry, Nicholas L. 1974. Knowledge Management: A New Concern for Public Administration. *Public Administration Review* 34 (3): 189–196.

Hermerén, Göran. 1986. *Kunskapens pris: Forskningsetiska problem och principer i humaniora och samhällsvetenskap* (The price of knowledge: Problems in research ethics and principles within humanities and social sciences). Stockholm: Humanistisk-samhällsvetenskapliga forskningsrådet.

Hess, David. 1997. *Science Studies: An Advanced Introduction.* New York: New York University Press.

Hesse, Bradford W., David E. Nelson, Gary L. Kreps, Robert T. Croyle, Neeraj K. Arora, Barbara K. Rimer, and Kasisomayajula Viswanath. 2005. Trust and Sources of Health Information: The Impact of the Internet and Its Implications for Health Care Providers: Findings from the First Health Information National Trends Survey. *Archives of Internal Medicine* 165 (22): 2618–2624.

Hesse-Biber, Sharlene Nagy, ed. 2011. *The Handbook of Emergent Technologies in Social Research.* Oxford: Oxford University Press.

Heuser, Ryan, and Long Le-Khac. 2011. Learning to Read Data: Bringing Out the Humanistic in the Digital Humanities. *Victorian Studies* 54 (1): 79–86.

Hey, Tony. 2004. Why Engage in e-Science? *CILIP Update* 3 (3): 25–27.

Hey, Tony, and Anne Trefethen. 2003. The Data Deluge: An e-Science Perspective. In *Grid Computing: Making the Global Infrastructure a Reality,* ed. Fran Berman, Geoffrey Fox, and Tony Hey, 809–824. Hoboken, NJ: Wiley.

Hey, Tony, and Anne Trefethen. 2006. Collaborative Physics, e-Science, and the Grid: Realizing Licklider's Dream. In *The New Physics for the Twenty-First Century*, ed. Gordon Fraser, 370–402. Cambridge: Cambridge University Press.

Hill, Kashmir. 2012. How Target Figured Out a Teen Girl Was Pregnant Before Her Father Did. Last modified February 16. http://www.forbes.com/sites/kashmirhill/2012/02/16/how-target-figured-out-a-teen-girl-was-pregnant-before-her-father-did/. Accessed February 20, 2012.

Hindman, Matthew. 2009. *The Myth of Digital Democracy*. Princeton, NJ: Princeton University Press.

Hine, Christine. 2006. Computerization Movements and Scientific Disciplines: The Reflexive Potential of New Technologies. In *New Infrastructures for Knowledge Production: Understanding e-Science*, ed. Christine Hine, 26–47. London: Information Science.

Hope, Janet. 2008. *Biobazaar: The Open Source Revolution and Biotechnology*. Cambridge, MA: Harvard University Press.

Houghton, John, and Peter Sheehan. 2006. *The Economic Impact of Enhanced Access to Research Findings*. Centre for Strategic Economic Studies (CSES) Working Paper no. 23. Melbourne, Australia: CSES, Victoria University.

Howard, Sharon, Tim Hitchcock, and Robert Shoemaker. 2010. *Crime in the Community Impact Analysis Report*. London: JISC.

Howard, Sharon, Tim Hitchcock, and Robert Shoemaker. 2011. *Crime in the Community: Enhancing User Engagement for Teaching and Research with the Old Bailey Online (Final Report)*. London: JISC.

Hughes, Jason, ed. 2012. *Sage Internet Research Methods*. Thousand Oaks, CA: Sage.

Hughes, Thomas P. 1994. Technological Momentum. In *Does Technology Drive History? The Dilemma of Technological Determinism*, ed. Leo Marx and Merritt Roe Smith, 101–113. Cambridge, MA: MIT Press.

Hunsinger, Jeremy, Lisbeth Klastrup, and Matthew Allen, eds. 2010. *International Handbook of Internet Research*. Dordrecht, Netherlands: Springer.

Iacono, Suzanne, and Rob Kling. 2001. Computerization Movements: The Rise of the Internet and Distant Forms of Work. In *Information Technology and Organizational Transformation: History, Rhetoric, and Practice*, ed. Joanne A. Yates and John Van Maanen, 93–136. Thousand Oaks, CA: Sage.

Jackson, Steven J., Paul N. Edwards, Geoffrey C. Bowker, and Cory P. Knobel. 2007. Understanding Infrastructure: History, Heuristics, and Cyberinfrastructure Policy. *First Monday* 12 (6).

Jankowski, Nicholas, ed. 2009. *e-Research: Transformations in Scholarly Practice*. London: Routledge.

Jasco, Peter. 2005. Google Scholar: The Pros and the Cons. *Online Information Review* 29 (2): 208–214.

Jockers, Matthew. 2013. *Macroanalysis: Digital Methods and Literary History*. Urbana: University of Illinois Press.

Jonsson, Lena, and Ulf Landegren. 2001. Storing and Using Biobanks for Research. In *The Use of Human Biobanks: Ethical, Social, Economical, and Legal Aspects*, ed. Mats G. Hansson, 1–8. Uppsala, Sweden: Uppsala University.

Justitiedepartementet (Sweden). 2007. Skyddet för den personliga integriteten: Kartläggning och analys (The protection of personal integrity: Mapping and analysis). *Statens offentliga utredningar* (SOU, Swedish government official reports) (2007): 22. http://www.regeringen.se/sb/d/8586/a/79592. Accessed 15 October 2009.

Keck, Margaret E., and Kathryn Sikkink. 1998. *Activists beyond Borders: Advocacy Networks in International Politics*. Ithaca, NY: Cornell University Press.

Kling, Rob. 1991. Computerization and Social Transformations. *Science, Technology & Human Values* 16 (3): 342–367.

Kling, Rob. 1992a. Audiences, Narratives, and Human Values in Social Studies of Technology. *Science, Technology & Human Values* 17 (3): 349–365.

Kling, Rob. 1992b. Behind the Terminal: The Critical Role of Computing Infrastructure in Effective Information Systems' Development and Use. In *Challenges and Strategies for Research in Systems Development*, ed. William Cotterman and James Senn, 153–201. London: Wiley.

Kling, Rob. 1999. What Is Social Informatics and Why Does it Matter? *D-Lib Magazine* 5 (1).

Kling, Rob, and Suzanne Iacono. 1988. The Mobilization of Support for Computerization: The Role of Computerization Movements. *Social Problems* 35 (3): 226–243.

Kling, Rob, and Suzanne Iacono. 1995. Computerization Movements and the Mobilization of Support for Computerization. In *Ecologies of Knowledge*, ed. Leigh Starr, 119–153. Albany: State University of New York Press.

Kling, Rob, and Geoffrey McKim. 2000. Not Just a Matter of Time: Field Differences and the Shaping of Electronic Media in Supporting Scientific Communication. *Journal of the American Society for Information Science* 51 (14): 1306–1320.

Kling, Rob, Geoffrey McKim, and Adam King. 2003. A Bit More to IT: Scholarly Communication Forums as Socio-technical Interaction Networks. *Journal of the American Society for Information Science and Technology* 54 (1): 46–67.

Kling, Rob, Howard Rosenbaum, and Steve Sawyer. 2005. *Understanding and Communicating Social Informatics: A Framework for Studying and Teaching the Human Contexts of Information and Communication Technologies*. Medford, NJ: Information Today.

Kling, Rob, and Walt Scacchi. 1982. The Web of Computing: Computer Technology as Social Organization. *Advances in Computers* 21:1–90.

Knorr Cetina, Karin. 1999. *Epistemic Cultures: How the Sciences Make Knowledge*. Cambridge, MA: Harvard University Press.

Kousha, Kayvan, and Mike Thelwall. 2007. Google Scholar Citations and Google Web/URL Citations: A Multi-discipline Exploratory Analysis. *Journal of the American Society for Information Science and Technology* 58 (7): 1055–1065.

Kuhn, Thomas. 1962. *The Structure of Scientific Revolutions*. Chicago: University of Chicago Press.

Kusch, Martin. 2010. Hacking's Historical Epistemology: A Critique of Styles of Reasoning. *Studies in History and Philosophy of Science* 41 (2): 158–173.

Kwa, Chunglin. 2011. *Styles of Knowing: A New History of Science from Ancient Times to the Present*. Pittsburgh: University of Pittsburgh Press.

Lamb, Roberta. 2006. Keynote speech at IFIP-TC9 World Conference on Human Choice and Computers, Maribor, Slovenia, September 21–23.

Lamb, Roberta, and Rob Kling. 2003. Reconceptualizing Users as Social Actors in Information Systems Research. *Management Information Systems Quarterly* 27 (2): 197–235.

Lamb, Roberta, Steve Sawyer, and Rob Kling. 2000. A Social Informatics Perspective on Socio-technical Networks. In *Proceedings of the Americas Conference on Information Systems*, ed. H. Michael Chung, paper 1. Atlanta: Association for Information Systems.

Latour, Bruno. 1987. *Science in Action: How to Follow Scientists and Engineers through Society*. Cambridge, MA: Harvard University Press.

Latour, Bruno, and Steve Woolgar. 1979. *Laboratory Life: The Social Construction of Scientific Facts*. Beverly Hills, CA: Sage.

Law, John, and John Hassard. 1999. *Actor Network Theory and After*. Malden, MA: Blackwell.

Lee, Raymond M. 2004. Recording Technologies and the Interview in Sociology, 1920–2000. *Sociology* 38 (5): 869–889.

Lemson, Gerard, and Jörg Colberg. 2004. Theory in the VO. http://wiki.ivoa.net/internal/IVOA/IvoaExecMeetingFM9/ivoa-20040129-theory.pdf. Accessed August 11, 2014.

Lenoir, Timothy, and Eric Giannella. 2006. The Emergence and Diffusion of DNA Microarray Technology. *Journal of Biomedical Discovery and Collaboration* 1: article 11.

Leydesdorff, Loet, Félix de Moya-Anegón, and Vicente P. Guerrero-Bote. 2014. Journal Maps, Interactive Overlays, and the Measurement of Interdisciplinarity on the Basis of Scopus Data (1996–2012). http://arxiv.org/abs/1310.4966. Accessed March 17, 2014.

Leydesdorff, Loet, Ismael Rafols, and Chaomei Chen. 2013. Interactive Overlays of Journals and the Measurement of Interdisciplinarity on the Basis of Aggregated Journal–Journal Citations. *Journal of the American Society for Information Science and Technology* 64 (12): 2573–2586.

Lichtenstein, P., U. De Faire, B. Floderus, M. Svartengren, P. Svedberg, and N. L. Pedersen. 2002. The Swedish Twin Registry: A Unique Resource for Clinical, Epidemiological, and Genetic Studies. *Journal of Internal Medicine* 252 (3): 184–205.

Lintott, Chris J., Kevin Schawinski, William Keel, Hanny Van Arkel, Nicola Bennert, Edward Edmondson, Daniel Thomas, et al. 2009. Galaxy Zoo: "Hanny's Voorwerp," a Quasar Light Echo? *Monthly Notices of the Royal Astronomical Society* 399 (1): 129–140.

Lintott, Chris J., Kevin Schawinski, Anže Slosar, Kate Land, Steven Bamford, Daniel Thomas, M. Jordan Raddick, et al. 2008. Galaxy Zoo: Morphologies Derived from Visual Inspection of Galaxies from the Sloan Digital Sky Survey. *Monthly Notices of the Royal Astronomical Society* 389 (3): 1179–1189.

MacKenzie, Donald. 1999. The Certainty Trough. In *Society on the Line*, ed. William H. Dutton, 43–46. Oxford: Oxford University Press.

Madrigal, Alexis. 2008a. Google Shutters Its Science Data Service. *Wired Science Blog*, http://www.wired.com/wiredscience/2008/12/googlescienceda/. Accessed December 19, 2008.

Madrigal, Alexis. 2008b. Google to Host Terabytes of Open-Source Science Data. *Wired Science Blog*, http://blog.wired.com/wiredscience/2008/01/google-to-provi.html. Accessed December 19, 2008.

Madsen, Christine. 2010. *Communities, Innovation, and Critical Mass: Understanding the Impact of Digitization on Scholarship in the Humanities through Tibetan and Himalayan Studies*. D.Phil. thesis, University of Oxford.

Mann, Michael. 2013. *The Sources of Social Power. Volume 4: Globalizations, 1945–2011*. New York: Cambridge University Press.

Manolio, Teri A., Laura Lyman Rodriguez, Lisa Brooks, Collaborative Association Study of Psoriasis, Dennis Ballinger, Mark Daly, Peter Donnelly, et al. 2007. New Models of Collaboration in Genome-wide Association Studies: The Genetic Association Information Network. *Nature Genetics* 39 (9): 1045–1051.

Marsh, Joanne, and Gill Evans. 2010. *D-TRACES Rapid Analysis Report*. London: JISC.

Matzat, Uwe. 2004. Academic Communication and Internet Discussion Groups: Transfer of Information or Creation of Social Contacts? *Social Networks* 26 (3): 221–255.

Mead, Clifford. 1989. *Thomas Pynchon: A Bibliography of Primary and Secondary Materials*. Elmwood Park, IL: Dalkey Archive Press.

Meho, Lokman I., and Helen Tibbo. 2003. Modelling the Information-Seeking Behaviour of Social Scientists: Ellis's Study Revisited. *Journal of the American Society for Information Science and Technology* 54 (6): 570–587.

Meho, Lokman I., and Kiduk Yang. 2007. Impact of Data Sources on Citation Counts and Rankings of LIS Faculty: Web of Science versus Scopus and Google Scholar. *Journal of the American Society for Information Science and Technology* 58 (13): 2105–2125.

Merton, Robert K. 1968. The Matthew Effect in Science. *Science* 159 (3810): 56–63.

Merton, Robert K. 1973. *The Sociology of Science*. Chicago: University of Chicago Press.

Meyer, Eric T. 2006. Socio-technical Interaction Networks: A Discussion of the Strengths, Weaknesses, and Future of Kling's STIN Model. In *Social Informatics: An Information Society for All?*, edited by Jacques Berleur, Marco I. Numinem, and John Impagliazzo, 37–48. International Federation for Information Processing, Volume 223. Boston: Springer.

Meyer, Eric T. 2007a. *Socio-technical Perspectives on Digital Photography: Scientific Digital Photography Use by Marine Mammal Researchers*. Ph.D. diss., Indiana University, Bloomington.

Meyer, Eric T. 2007b. Technological Change and the Form of Science Research Teams: Dealing with the Digitals. *Prometheus* 25 (4): 345–361.

Meyer, Eric T. 2008. Framing the Photographs: Understanding Digital Photography as a Computerization Movement. In *Computerization Movements and Technology Diffusion: From Mainframes to Ubiquitous Computing*, ed. Ken L. Kraemer and Margaret S. Elliott, 173–199. Medford, NJ: Information Today.

Meyer, Eric T. 2009. Moving from Small Science to Big Science: Social and Organizational Impediments to Large Scale Data Sharing. In *e-Research: Transformation in Scholarly Practice*, ed. Nicholas Jankowski, 147–159. New York: Routledge.

Meyer, Eric T. 2011. *Splashes and Ripples: Synthesizing the Evidence on the Impact of Digital Resources*. London: JISC.

Meyer, Eric T. 2014. Examining the Hyphen: The Value of Social Informatics for Research and Teaching. In *Social Informatics: Past, Present and Future*, ed. Pnina

Fichman and Howard Rosenbaum, 57–74. Cambridge, UK: Cambridge Scholarly Publishers.

Meyer, Eric T., Monica Bulger, Avgousta Kyriakidou-Zacharoudiou, Lucy Power, Peter Williams, Will Venters, Melissa Terras, and Sally Wyatt. 2012. *Collaborative yet Independent: Information Practices in the Physical Sciences*. London: Research Information Network.

Meyer, Eric T., and William H. Dutton. 2009. Top-Down e-Infrastructure Meets Bottom-Up Research Innovation: The Social Shaping of e-Research. *Prometheus* 27 (3): 239–250.

Meyer, Eric T., Kathryn Eccles, Michael Thelwall, and Christine Madsen. 2009. *Final Report to JISC on the Usage and Impact Study of JISC-Funded Phase 1 Digitisation Projects & the Toolkit for the Impact of Digitised Scholarly Resources (TIDSR)*. London: JISC.

Meyer, Eric T., and Isis A. Hjorth. 2013. *Digitally Scratching New Theatre: London's Battersea Arts Centre Engaging via the Web*. London: Nesta.

Meyer, Eric T., and Rob Kling. 2002. *Leveling the Playing Field, or Expanding the Bleachers? Socio-technical Interaction Networks and arXiv.org*. Center for Social Informatics Working Paper Series WP-02-10. http://www.slis.indiana.edu/CSI/WP/WP02-10B.html. Accessed April 2, 2002.

Meyer, Eric T., Anne-Marie Oostveen, Ralph Schroeder, Michael Boniface, J. Brian Pickering, Paul Walland, Burkhard Stiller, and Martin Waldburger. 2012. *Final Report on Social Future Internet Coordination Activities*. Zurich: SESERV Consortium.

Meyer, Eric T., and Ralph Schroeder. 2009a. Untangling the Web of e-Research: Towards a Sociology of Online Knowledge. *Informetrics* 3 (3): 246–260.

Meyer, Eric T., and Ralph Schroeder. 2009b. The World Wide Web of Research and Access to Knowledge. *Journal of Knowledge Management Research and Practice* 7 (3): 218–233.

Meyer, Eric T., and Ralph Schroeder. 2013. Digital Transformations of Scholarship and Knowledge. In *The Oxford Handbook of Internet Studies*, ed. William H. Dutton, 307–327. Oxford: Oxford University Press.

Meyer, Eric T., Ralph Schroeder, and William H. Dutton. 2008. The Role of e-Infrastructures in the Transformation of Research Practices and Outcomes. Paper presented at the iConference 2008, UCLA, Los Angeles, CA, February 28–March 1.

Meyer, Eric T., Arthur Thomas, and Ralph Schroeder. 2011. *Web Archives: The Future(s)*. London: International Internet Preservation Consortium.

Michel, Jean-Baptiste, Yuan Kui Shen, Aviva Presser Aiden, Adrian Veres, Matthew K. Gray, Google Books Team, Joseph P. Pickett, et al. 2011. Quantitative Analysis of Culture Using Millions of Digitized Books. *Science* 331 (6014): 176–182.

Mons, Barend, Michael Ashburner, Christine Chichester, Erik Van Mulligen, Marc Weeber, Johan Den Dunnen, Gert-Jan Van Ommen, et al. 2008. Calling on a Million Minds for Community Annotation in *WikiProteins*. *Genome Biology* 9 (5): R89.1–R89.15.

Moretti, Franco. 2000. Conjectures on World Literature. *New Left Review* 1 (January–February): 54–68.

Moretti, Franco. 2005. *Graphs, Maps, Trees: Abstract Models for a Literary History*. London: Verso Books.

Moretti, Franco. 2011. Network Theory, Plot Analysis. *New Left Review* 68:80–102.

Moretti, Franco. 2013. *Distant Reading*. London: Verso Press.

Moschovakis, Yiannis N. 2001. What Is an Algorithm? In *Mathematics Unlimited—2001 and Beyond*, ed. Björn Engquist and Wilfried Schmid, 919–936. Berlin: Springer.

Mumford, Michael D., Ginamarie M. Scott, Blaine Gaddis, and Jill M. Strange. 2002. Leading Creative People: Orchestrating Expertise and Relationships. *Leadership Quarterly* 13 (6): 705–750.

Negroponte, Nicholas. 1995. *being digital*. New York: Knopf.

Nentwich, Michael, and René König. 2012. *Cyberscience 2.0: Research in the Age of Digital Social Networks*. Frankfurt: Campus.

Newton, Isaac. 1959. *The Correspondence of Isaac Newton, I: 1661–1675*. Edited by H. W. Turnbull. Cambridge: Cambridge University Press for the Royal Society.

Nicholas, David, Paul Huntington, Hamid R. Jamali, and Anthony Watkinson. 2006. The Information Seeking Behaviour of the Users of Digital Scholarly Journals. *Information Processing & Management* 42 (5): 1345–1365.

Nunberg, Geoffrey. 2009. Google's Book Search: A Disaster for Scholars. *Chronicle Review* 31. http://chronicle.com/article/Googles-Book-Search-A/48245/. Accessed August 24, 2010.

Olson, Gary M., Ann Zimmerman, and Nathan D. Bos, eds. 2008. *Scientific Collaboration on the Internet*. Cambridge, MA: MIT Press.

Ordnance Survey. 2008. Overview. http://www.ordnancesurvey.co.uk/oswebsite/aboutus/yourinforights/. Accessed January 23, 2008.

Orlikowski, Wanda J. 2000. Using Technology and Constituting Structures: A Practice Lens for Studying Technology in Organizations. *Organization Science* 11 (4): 404–428.

Otjacques, Benoit, Patrik Hitzelberger, and Fernand Feltz. 2006. Identity Management and Data Sharing in the European Union. In *Proceedings of the 39th Hawaii*

References

International Conference on System Sciences, ed. Ralph H. Sprague, Jr., paper 70. Piscataway, NJ: IEEE Computer Society.

Oudshoorn, Nelly, and Trevor Pinch, eds. 2005. *How Users Matter: The Co-construction of Users and Technology*. Cambridge, MA: MIT Press.

Pakhira, Anjan, Ronald Fowler, Lakshmi Sastry, and Toby Perring. 2005. Grid Enabling Legacy Applications for Scalability—Experiences of a Production Application on the UK NGS. Paper presented at the UK e-Science All Hands Meeting (AHM'05), Nottingham, UK, September 19–22.

Pannapacker, William. 2009. The MLA and the Digital Humanities. *Chronicle of Higher Education Blog*, December 28. http://chronicle.com/blogPost/The-MLAthe-Digital/19468/. Accessed 27 January 2010.

Pannapacker, William. 2011. Pannapacker at MLA: Digital Humanities Triumphant? *Chronicle of Higher Education Blog*, January 8. http://chronicle.com/blogs/brainstorm/pannapacker-at-mla-digital-humanities-triumphant/30915. Accessed January 9, 2011.

Patterson, Troy. 2006. The Pynchon Post: Did the Master Make an Appearance on His Amazon Page? *Slate*, July 19. http://www.slate.com/articles/arts/culturebox/2006/07/the_pynchon_post.html. Accessed January 6, 2011.

Pearce, Sarah E., and Will Venters. 2013. How Particle Physicists Constructed the World's Largest Grid: A Case Study in Participative Cultures. In *The Routledge Handbook of Participatory Cultures*, ed. Aaron Delwiche and Jennifer Henderson, 130–140. New York: Routledge.

Pinch, Trevor J., and Wiebe E. Bijker. 1987. The Social Construction of Facts and Artifacts: Or How the Sociology of Science and the Sociology of Technology Might Benefit Each Other. In *The Social Construction of Technological Systems*, ed. Wiebe E. Bijker, Thomas P. Hughes, and Trevor Pinch, 17–50. Cambridge, MA: MIT Press.

Podvinec, Michael, Sergio Maffioletti, Peter Kunszt, Konstantin Arnold, Lorenzo Cerutti, Bruno Nyffeler, Ralph Schlapbach, et al. 2006. The SwissBioGrid Project: Objectives, Preliminary Results and Lessons Learned. In *Proceedings of e-Science 2006: Second IEEE International Conference on e-Science and Grid Computing*, paper 148. Piscataway, NJ: IEEE.

Power, Lucy. 2011. *e-Research in the Life Sciences: From Invisible to Virtual Colleges*. D.Phil. thesis, University of Oxford.

Pynchon, Thomas. 2000. *Slow Learner: Early Stories*. London: Vintage.

Pynchon, Thomas. 2006. *Against the Day*. New York: Penguin.

Pynchon Wiki. 2007. Thomas Pynchon. http://against-the-day.pynchonwiki.com/wiki/index.php?title=Thomas_Pynchon. Accessed October 31, 2007.

Raddick, M. Jordan, Georgia Bracey, Pamela L. Gay, Chris J. Lintott, Phil Murray, Kevin Schawinski, Alexander S. Szalay, and Jan Vandenberg. 2010. Galaxy Zoo: Exploring the Motivations of Citizen Science Volunteers. *Astronomy Education Review* 9 (1).

Ramachandran, Rahul, Sara Graves, Helen Conover, and Karen Moe. 2004. Earth Science Markup Language (ESML): A Solution for the Scientific Data-Application Interoperability Problem. *Computers & Geosciences* 30 (1): 117–124.

Reagle, Joseph M. 2010. *Good Faith Collaboration: The Culture of Wikipedia*. Cambridge, MA: MIT Press.

Restivo, Sal, and Jennifer Croissant. 2008. Social Constructionism in Science and Technology Studies. In *Handbook of Constructionist Research*, ed. James A. Holstein and Jaber F. Gubrium, 213–229. New York: Guilford Press.

Rieger, Oya. 2009. Search Engine Use Behaviour of Students and Faculty: User Perceptions and Implications for Future Research. *First Monday* 14 (12).

Rogers, Everett. 2003. *Diffusion of Innovations*. 5th ed. New York: Free Press.

Rowberry, Simon Peter. 2012. Reassessing the *Gravity's Rainbow Pynchon Wiki*: A New Research Paradigm? *Orbit: Writing around Pynchon* 1 (1).

Rushkoff, Douglas. 1996. *Playing the Future*. New York: HarperCollins.

Salmon, Felix, and Jan Stoke. 2010. Algorithms Take Control of Wall Street. *Wired*, December 27.

Sandstrom, Pamela. 2001. Scholarly Communication as a Socioecological System. *Scientometrics* 51 (3): 573–605.

Sathe, Nila A., Jenifer L. Grady, and Nunzia B. Giuse. 2002. Print versus Electronic Journals: A Preliminary Investigation into the Effect of Journal Format on Research Processes. *Journal of the Medical Library Association: JMLA* 90 (2): 235–243.

Savage, Mike, and Roger Burrows. 2007. The Coming Crisis of Empirical Sociology. *Sociology* 41 (5): 885–899.

Savage, Mike, and Roger Burrows. 2009. Some Further Reflections on the Coming Crisis of Empirical Sociology. *Sociology* 43 (4): 762–772.

Scacchi, Walt. 2005. Socio-technical Interaction Networks in Free/Open Source Software Development Processes. In *Software Process Modeling*, ed. Silvia T. Acuña and Natalia Juristo, 1–27. New York: Springer Science+Business Media.

Schneider, Steven M., and Kirsten A. Foot. 2005. Web Sphere Analysis: An Approach to Studying Online Action. In *Virtual Methods: Issues in Social Research on the Internet*, ed. Christine Hine, 157–170. Oxford: Berg.

Schroeder, Ralph. 2006. e-Sciences: Infrastructures That Reshape the Global Contours of Knowledge. Paper presented at the Second International Conference on e–Social Science, Manchester, UK, June 28–30.

Schroeder, Ralph. 2007a. e-Research Infrastructures and Open Science: Towards a New System of Knowledge Production? *Prometheus* 25 (1): 1–17.

Schroeder, Ralph. 2007b. *Rethinking Science, Technology, and Social Change.* Stanford, CA: Stanford University Press.

Schroeder, Ralph. 2008. e-Sciences as Research Technologies: Reconfiguring Disciplines, Globalizing Knowledge. *Social Sciences Information / Information sur les sciences sociales* 47 (2): 131–157.

Schroeder, Ralph. 2013a. *An Age of Limits: Social Theory for the 21st Century.* Basingstoke, UK: Palgrave Macmillan.

Schroeder, Ralph. 2013b. Big Data and the Uses and Disadvantages of Scientificity for Social Research. Paper presented at the American Sociological Association Annual Meeting, New York, August 10–13.

Schroeder, Ralph. 2014. Big Data: Towards a More Scientific Social Science and Humanities? In *Society and the Internet*, ed. Mark Graham and William H. Dutton, 164–176. Oxford: Oxford University Press.

Schroeder, Ralph, and Matthijs den Besten. 2008. Literary Sleuths Online: e-Research Collaboration on the *Pynchon Wiki*. *Information Communication and Society* 11 (2): 167–187.

Schroeder, Ralph, Matthijs den Besten, and Jenny Fry. 2007. Catching Up or Latecomer Advantage: Lessons from e-Research Strategies in Germany, in the UK and Beyond. Paper presented at the German e-Science Conference 2007, Baden-Baden, May 2–4.

Schroeder, Ralph, and Jenny Fry. 2007. Social Science Approaches to e-Science: Framing an Agenda. *Journal of Computer-Mediated Communication* 12 (2): 563–582.

Schroeder, Ralph, and Eric T. Meyer. 2012. Digital Research and Big Data: Is the Tail Wagging the Dog? Paper presented at the Digital Research 2012 Conference, Oxford, September 10–12.

Schroeder, Ralph, and Dimitrina Spencer. 2009. Social Scientists and the Domestication of e-Research Tools. Paper presented at the 5th International Conference on e–Social Science, Cologne, Germany, June 24–26.

Segev, Elad, and Niv Ahituv. 2010. Popular Searches in Google and Yahoo! A "Digital Divide" in Information Uses? *Information Society* 26 (1): 17–37.

Select Committee on Science and Technology. 2002. *What on Earth? The Threat to the Science Underpinning Conservation.* Science and Technology, Third Report. London: House of Lords.

Sell, Susan K. 2003. *Private Power, Public Law: The Globalization of Intellectual Property Rights.* Cambridge, UK: Cambridge University Press.

Sellen, Abigail, and Richard H. R. Harper. 2001. *The Myth of the Paperless Office.* Cambridge, MA: MIT Press.

Shimojo, Fuyuki, Rajiv K. Kalia, Aiichiro Nakano, and Priya Vashishta. 2001. Linear-Scaling Density-Functional-Theory Calculations of Electronic Structure Based on Real-Space Grids: Design, Analysis, and Scalability Test of Parallel Algorithms. *Computer Physics Communications* 140 (3): 303–314.

Shinn, Terry. 2008. *Research Technology and Cultural Change: Instrumentation, Genericity, Transversality.* Oxford: Bardwell Press.

Shinn, Terry, and Bernward Joerges. 2002. The Transverse Science and Technology Culture: Dynamics and Roles of Research–Technology. *Social Sciences Information / Information sur les sciences sociales* 41 (2): 207–251.

Shrum, Wesley, Joel Genuth, and Ivan Chompalov. 2007. *Structures of Scientific Collaboration.* Cambridge, MA: MIT Press.

Siefring, Judith, and Eric T. Meyer. 2013. *Sustaining the EEBO–TCP Corpus in Transition: Report on the TIDSR Benchmarking Study.* London: JISC.

Simmhan, Yogesh L., Beth Plale, and Dennis Gannon. 2005. A Survey of Data Provenance in e-Science. *SIGMOD Record* 34 (5): 31–36.

Slaughter, Anne-Marie. 2004. *A New World Order.* Princeton, NJ: Princeton University Press.

Statistics Sweden. 2012. Webpage, English version. http://www.scb.se/en_/. Accessed July 19, 2012.

Stauffer, Andrew. 2011. Introduction: Searching Engines, Reading Machines. *Victorian Studies* 54 (1): 63–68.

Stenberg, S. A., and D. Vågerö. 2006. Cohort Profile: The Stockholm Birth Cohort of 1953. *International Journal of Epidemiology* 35 (3): 546–548.

Stokes, Donald E. 1997. *Pasteur's Quadrant: Basic Science and Technological Innovation.* Washington, DC: Brookings Institution.

Tancer, Bill. 2009. *Click: What We Do Online and Why It Matters.* London: HarperCollins UK.

Tanner, Simon. 2010. *Inspiring Research, Inspiring Scholarship.* London: JISC.

Tanner, Simon, and Marilyn Deegan. 2011. *Inspiring Research, Inspiring Scholarship: The Value and Benefits of Digitised Resources for Learning, Teaching, Research, and Enjoyment.* London: JISC.

Tatum, Clifford, and Michelle LaFrance. 2009. Wikipedia as Distributed Knowledge Laboratory: The Case of Neoliberalism. In *e-Research: Transformation in Scholarly Practice*, ed. Nicholas Jankowski, 310–327. New York: Routledge.

Tenopir, Carol, Donald W. King, Peter Boyce, Matt Grayson, and Keri-Lynn Paulson. 2005. Relying on Electronic Journals: Reading Patterns of Astronomers. *Journal of the American Society for Information Science and Technology* 56 (8): 786–802.

Tenopir, Carol, Donald W. King, and Amy Bush. 2004. Medical Faculty's Use of Print and Electronic Journals: Changes over Time and in Comparison with Scientists. *Journal of the Medical Library Association: JMLA* 92 (2): 233–241.

Tenopir, Carol, and Rachel Volentine. 2012. *UK Scholarly Reading and the Value of Library Resources: Summary Results of the Study Conducted Spring 2011.* London: JISC Collections.

Thelwall, Mike. 2009. *Introduction to Webometrics: Quantitative Web Research for the Social Sciences.* Ed. Gary Marchionini. Synthesis Lectures on Information Concepts, Retrieval, and Services, vol. 1, no. 1. San Rafael, CA: Morgan & Claypool.

Thelwall, Mike, and Kayvan Kousha. 2008. Online Presentations as a Source of Scientific Impact? An Analysis of PowerPoint Files Citing Academic Journals. *Journal of the American Society for Information Science and Technology* 59 (5): 805–815.

Thomas, Arthur, Eric T. Meyer, Meghan Dougherty, Charles Van den Heuvel, Christine Madsen, and Sally Wyatt. 2010. *Researcher Engagement with Web Archives: Challenges and Opportunities for Investment.* London: JISC.

US National Institutes of Health (NIH). 2003. NIH Data Sharing Policy and Implementation Guidance. Last modified March 5. http://grants.nih.gov/grants/policy/data_sharing/data_sharing_guidance.htm. Accessed July 19, 2012.

Vallas, S. P., and D. L. Kleinman. 2008. Contradiction, Convergence, and the Knowledge Economy: The Confluence of Academic and Commercial Biotechnology. *Socio-economic Review* 6 (2): 283–311.

Van de Sompel, Herbert, Sandy Payette, John Erickson, Carl Lagoze, and Simeon Warner. 2004. Rethinking Scholarly Communication. *D-Lib Magazine* 10 (9).

Van der Vleuten, Erik. 2004. Infrastructures and Societal Change: A View from the Large Technical Systems Field. *Technology Analysis and Strategic Management* 16 (3): 395–414.

Vaughan, Jason. 2010. Insights into the Commons on Flickr. *Libraries and the Academy* 10 (2): 185–214.

Vehovar, Vasja, and Katja Lozar Manfreda. 2008. Overview: Online Surveys. In *The Sage Handbook of Online Research Methods*, ed. Nigel Fielding, Raymond M. Lee, and Grant Blank, 177–194. Thousand Oaks, CA: Sage.

Vetenskapsrådet. 2005. *Strategi och infrastruktur för världsledande forskning på svenska register* (Strategy and infrastructure for world leading research on Swedish registers). Internal report based on an investigation concerning the possibilities of improving the conditions for research on Swedish register data. Stockholm: Vetenskapsrådet.

Vetenskapsrådet. 2009. Web page. http://www.vr.se/. Accessed March 18, 2009.

Vickers, Andrew J. 2006. Whose Data Set Is It Anyway? Sharing Raw Data from Randomized Trials. *Trials* 7 (1): article 15.

Wagner, Caroline S., and Loet Leydesdorff. 2005. Network Structure, Self-Organization, and the Growth of International Collaboration in Science. *Research Policy* 34 (10): 1608–1618.

Waller, Vivienne. 2009. The Relationship between Public Libraries and Google: Too Much Information. *First Monday* 14 (9).

Waller, Vivienne. 2011. Not Just Information: Who Searches for What on the Search Engine Google? *Journal of the American Society for Information Science and Technology* 62 (4): 761–775.

Walsh, John P., S. Kucker, N. G. Maloney, and S. Gabbay. 2000. Connecting Minds: Computer-Mediated Communication and Scientific Work. *Journal of the American Society for Information Science* 51 (14): 1295–1305.

Walsh, John P., and Nancy G. Maloney. 2007. Collaboration Structure, Communication Media, and Problems in Scientific Work Teams. *Journal of Computer-Mediated Communication* 12 (2): 712–732.

WebCite. 2010. Web page. http://www.Webcitation.org/. Accessed July 28, 2010.

Weber, Steven. 2004. *The Success of Open Source*. Cambridge, UK: Cambridge University Press.

Weisenburger, Steven C. 1988. *A Gravity's Rainbow Companion: Sources and Context for Pynchon's Novel*. Athens: University of Georgia Press.

Weisenburger, Steven C. 2006. *A Gravity's Rainbow Companion: Sources and Contexts for Pynchon's Novel*. 2nd ed. Athens: University of Georgia Press.

Welin, Stellan. 1990. The Computerized Scientist—Moral and Political Issues. In *In Science We Trust? Moral and Political Issues of Science and Society*, ed. Aant Elzinga, Jan Nolin, Rob Pranger, and Sune Sunesson, 155–176. Lund, Sweden: Lund University Press.

References

Whatley, Sarah, Andrea Barzey, Joanne Marsh, Gill Evans, Ross Varney, and John Tutchings. 2011. *D-TRACES Project (Dance Teaching Resource and Collaborative Engagement Spaces): Final Report*. London: JISC.

Whitley, Richard. 2000. *The Intellectual and Social Organization of the Sciences*. 2nd ed. Oxford: Oxford University Press.

Wilkinson, David, Gareth Harries, Mike Thelwall, and Liz Price. 2003. Motivations for Academic Web Site Interlinking: Evidence for the Web as a Novel Source of Information on Informal Scholarly Communication. *Journal of Information Science* 29 (1): 49–56.

Willinsky, John. 2005a. *The Access Principle: The Case for Open Access to Research and Scholarship*. Cambridge, MA: MIT Press.

Willinsky, John. 2005b. The Unacknowledged Convergence of Open Source, Open Access, and Open Science. *First Monday* 10 (8).

Wilson, Rowan, Carl Marshall, and Fawei Geng. 2010. *Listening for Impact Rapid Analysis Report*. London: JISC.

Wood, Michael. 2007. Humming Along (Review of *Against the Day*). *London Review of Books* 29 (1): 12–13.

Wuchty, Stefan, Benjamin F. Jones, and Brian Uzzi. 2007. The Increasing Dominance of Teams in Production of Knowledge. *Science* 316:1036–1039.

Yakel, Elizabeth. 2004. Seeking Information, Seeking Connections, Seeking Meaning: Genealogists and Family Historians. *Information Research* 10 (1).

Zheng, Yingqin, Will Venters, and Tony Cornford. 2011. Collective Agility, Paradox, and Organizational Improvisation: The Development of a Particle Physics Grid. *Information Systems Journal* 21 (4): 303–333.

Zins, Chaim. 2007. Conceptual Approaches for Defining Data, Information, and Knowledge. *Journal of the American Society for Information Science and Technology* 58 (4): 479–493.

Index

Note: italicized page numbers denote figures and tables.

Aarseth, Espen, 60
Academia-industry divide, challenges across, 14–15
Accelerating Transition to Virtual Research Organization in Social Science (AVROSS), 45, 66
Accidental e-researchers, 6, 126–128
Ackland, Rob, 140
Actor-network theory, 35, 36
  epistemic communities and, 180
Adhocracy, 80
Against the Day (Pynchon), 87–93
Aggregation. *See* Collaborative computing
American National Election Surveys, 33
Anderson, Chris, 170
Archives, web, 190–193
Arts and Humanities e-Science Support Centre, 47
Arts and Humanities Research Council, 147
ArXiv.org, 58, 164
Astronomical data, 134, 137–138, 212
Atkins report, 45, 46
Attention space, 181–184
Australian National University, 16
Autonomization of science and knowledge, 220

Axelsson, Anne-Sofie, 105

Barjak, Franz, 48, 165
Beatles, the, 17
Benkler, Yochai, 91, 185
Berman, Francine, 45, 46, 139
Bertenthal, Bennett, 47
Big data, 26, 49–51, 123
  social science, 143–145
Big science, 215
Bijker, Wiebe, 36, 39
Bodleian Libraries, 55
Borgman, Christine, 95, 163, 165, 167
Boundaries of e-research, 6
Brady, Henry, 45, 46, 139
British History Online, 147, 156
British Household Panel Survey, 33
Burrows, Roger, 5, 145

Callon, Michel, 36
Cameron, Graham, 78–79
Caro, Robert, 204
Carusi, Annamaria, 194
Center for Embedded Network Sensing (CENS), 11, 118
Citizen science, 2, 10–11, 82–84, *200*

Cloud computing, 26, 50, 80, 82, 160, 175, 210, 212
Coauthorship, 60–63
Collaborative computing. *See also* Data sharing
  aggregating machines, 72–81
  aggregating many eyes for science, 81–86
  aggregating many minds for the humanities, 86–93
  complexity continuum and, 69–72
  data sharing and, 114–115
  disciplinary differences in, 207–208
  GAIN, 101–105
  scalability and, 96
  SPLASH, 15, 97–101
Collaboratories, 46
Collins, Randall, 25–30, 203
Communism, 177, 178, 186
Complexities of data sharing, 11–14
Complexity continuum, 69–72, 94
Computerization movements (CMs), 38–41, 210, 211–216, 221
Computer science and engineering, disciplinary differences in e-research and, 128–134
Construction of analogical models. *See* Modeling
Council of European Social Science Data Archives, 139
Creative Commons licensing, 16, 140, 176, 233n5
Crombie, A. C., 23–24
Crowdsourcing, 16
Cultures of data sharing, 95–97
Culturomics, 153–154
Cummings, Jonathon, 168
Cumulative sciences, 203–204
Cunningham, Ward, 86
Cyberinfrastructure, 5, 45, 52, 72
  accidental e-researchers and, 126
Cybermetrics, 142

Data sharing, 123–124. *See also* Distributed data
  complexities of, 11–14
  cultures of, 95–97
  decision making and, 112–113
  ethical issues in, 194
  limits of, 193–194
  privacy and trust in, 194–195
Davies, Siobhan, 149–151, 232n5
Desk research, digital transformations of, 157–172
Digital Curation Centre, 47
Digital humanities. *See* Humanities
Digital photography, 122–123
Disciplinary differences
  in e-research, 128–134, 168–169
  in transformation of research, 206–208
Disinterestedness, 177
Distant reading, 92, 153–154
Distributed data
  ad hoc requests, 113
  collaborations, 114–115
  cultures of data sharing and, 95–97
  data management policy and, 115
  data manipulation and, 118, 119
  data sharing in Sweden and, 105–112
  decision making and, 112–114
  marine mammal science, 97–101
  phenotypic data as, 121–122
  psychiatric genetics, 12–13, 101–105
  sensors and, 118–119
  as shift from small to big science, 112–124
  speed and organization of data in, 119–120

Early English Books Online (EEBO), 55–56, 155–156
Economic constraints on e-research, 195–196
Edwards, Paul, 236n7

Enabling Grids for E-sciencE (EGEE), 134, 135–137
Enclosure movement, 179
End users, 126
Epistemic communities, 180
e-Research. *See also* Limits of e-research
  approaches to understanding, 7–8
  boundaries of, 6
  challenges of infrastructure across academia-industry divide and, 14–15
  changing everyday practices and supporting researchers, 15–16
  coauthorship, 60–61
  complexities of data sharing and, 11–14
  data sharing in, 123–124
  defining, 4–7
  disciplinary differences in, 128–134, 168–169
  economic constraints on, 195–196
  engaging communities in the humanities, 16–18
  funding, 46–48
  geographic distribution of, 57–60, 61–63, *64*
  limitations of openness in, 181–184
  mutual dependence, 33
  networks enabling, 211–216
  new mechanisms enabling, 218–223
  norms of scientific inquiry and, 177
  output mapping, 48–67
  physics and large-scale data, 9–10
  publications related to, 52–67
  and public engagement via citizen science, 10–11
  scale and scope of knowledge production and, 124
  social informatics and, 34–38
  in society and academic disciplines, 41–43
  styles of science and, 23–34, 197–206
  task uncertainty, 33
  technology and regulations collide, 187–190
  top-down perspective in, 127
  transforming research, 8–18
  visibility and access, 18–19
  web archives, 190–193
e-Researchers
  accidental, 126–128
  definition of, 125–126
e-Scholarly communication layer (e-SCL), 170–171
e-Science, 5, 40–41, 45–48, 62, 106, 216
e-Social science, 5, 45–46
Ethical issues in data sharing, 194
Ethos of modern science, 177
European Bioinformatics Institute, 78–79
European Commission, 45, 63
European Grid Infrastructure (EGI), 134
European Organization for Nuclear Research (CERN), 2, 9
European Social Science Data Archives, 139
European Social Survey, 139
European Strategy Forum on Research Infrastructures, 139
Experimentation, 23–24

Facebook, 125, 145, 149, 204
Fischer, Michael, 66
Freeman, Linton, 32
Frickel, Scott, 41
Fry, Jenny, 167
Fuchs, Stephan, 178
Funding of e-research, 46–48

GAIN. *See* Genetic Association Information Network (GAIN)
Galaxy Zoo, 10, 11, 16, 17, 18, 81–86, 119, 124
  scientific style, 198, *200*, 205

Garvey, William, 164
Genetic Association Information
  Network (GAIN), 12–13, 101–105,
  127
  collaboration on, 207
  limits of, 193–194
  scientific style, *200*, 202, 205
  in shift from small to big science,
  112–124
Geographic distribution of e-research,
  57–60, 61–63, *64*
GeoVue, 187–190
German D-Grid initiative, 139
German Federal Ministry of Education,
  47
Gillberg, Christopher, 109, 110
Giuse, Nunzia, 161
Global Information Commons for
  Science Initiative, 177
Goertz, Gary, 204
Goodman, Alyssa, 157
Google, 1–2, 160, 165, 196
  big social science data and, 143–145
  Books project, 153, 154
  Earth, 188–189
  Generation, 162–163
  NGram Viewer, 37, 154
  Scholar, 18, 55, *56*, 169–170
  transparency and, 172
Grady, Jenifer, 161
Gravity's Rainbow (Pynchon), 87, 90, 92
Grid computing, 9, 28, 49–50, 72, 76,
  210, 225n7
Grid Particle Physics (GridPP), 9–10, 15
Griffith, Belver, 164
Gross, Neil, 41

Haas, Peter, 180
Hacker ethic, 185
Hackett, Edward, 80
Hacking, Ian, 23, 25, 27, 118, 197, 203,
  205, 206
Hallmark, Julie, 161

Hanny's Voorwerp, 10, 84, 119, 198
Hara, Noriko, 40
Harnad, Stevan, 168
Harvard Business Review, 164
Heimeriks, Gaston, 165
Heller, Agnes, 203
Hermerén, Göran, 109
Hess, David, 177
Heuser, Ryan, 154
High-consensus rapid-discovery science,
  26–28
Hine, Christine, 40
Historicogenetic explanation, 24
Histpop (the Online Historical
  Population Reports), 150, 232n8
Hjorth, Isis, 149
Hooke, Richard, 34
Hughes, Thomas, 28
Humanities
  aggregating many minds for,
    86–93
  collaboration in, 207–208
  as cumulative, 204
  digital resources available to scholars
    in, 147–149
  digital transformations of desk
    research in, 157–172
  digitized collections' impact on
    teaching in, 149–152
  disciplinary differences in e-research
    and, 128–134
  e-research engaging communities in,
    16–17
  e-research publications in, 54,
    166–167
  generational differences and
    e-research in, 161–163
  informal scientific communication in,
    163–165
  large-scale textual analysis, 152–154
  open-access publishing in, 165–166
  resistance to digital resources in,
    154–157

Iacono, Suzanne, 39–40
Informal scientific communication, 163–165
Information society, 8
Information source, the Internet as, 3–4
Infrastructures, e-research, 181–184
Institute for Scientific Information (ISI), 164, 169
Institute of Molecular Systems Biology, 74
Intellectual property rights, 179, 233–234n11
Interdisciplinarity, 206
International Council for Science, 177
International Virtual Observatory Alliance (IVOA), 134, 137–138
Internet research, 1–3
  about engineering and computer science, 4
  Internet as information source and, 3–4
  Internet as research tool and research method and, 4
  Internet as social phenomenon and, 3
  net neutrality and, 234n18
  typology of, 3–4
  underlying technologies of, 4
ITunes U, 150

Jirotka, Marina, 194
Joerges, Bernward, 28–30, 50, 76, 142
Johnson, Lyndon, 204
JSTOR, 55, 56

Kiesler, Sara, 168
King, Adam, 35, 36, 166
Kleinman, Daniel, 80, 81
Kling, Rob, 35, 36, 166, 167
  on computerization movements, 38–41
Knowledge
  automatization of, 220
  definition of, 6

external conditions of, 220
machinification of, 215
production, scale, and scope of, 124
variety and homogeneity in transformations of, 216–218
Knowledge machines
  CMs, SIMs, and networks enabling e-research and, 211–216
  disciplinary differences in transformation of research through, 206–208
  new mechanisms, 218–223
  scientific styles and, 197–206
  STINs and momentum of digital research and, 208–211
  variety and homogeneity in transformations of knowledge and, 216–218
Kousha, Kayvan, 169
Kuhn, Thomas, 19, 39, 203

Lamb, Robert, 36, 212
Lane, Julia, 48
Large Hadron Collider (LHC), 2, 9–10
  scientific style, 205
Large-scale textual analysis in digital humanities, 152–154
Latour, Bruno, 36
Law, John, 36
Le Khac, Long, 154
Li, Xuemei, 165
Life sciences, 78–81
  SPLASH project, 15, 97–101
Limits of e-research
  digitally dusty web archives and, 190–193
  economic constraints and, 195–196
  ethical issues and, 194
  limits of data sharing and, 193–194
  privacy and trust and, 194–195
  technology and regulations collide, 187–190

LinkedIn, 125
Linkrot, 156
Litlab, 153
Lock-ins, 181
Lyon, Stephen, 66

Machine aggregation, 72–81
Machinification of knowledge, 215
Macy, Michael, 47
Mahoney, James, 204
Maloney, Nancy, 168
Mapping e-research output, 48–67
 big data, 50–51
 Cloud computing and, 50
 supercomputing, 51–52
Marine mammal science, 15, 97–101. *See also* Structure of Populations, Levels of Abundance, and Status of Humpbacks (SPLASH)
Mathematical style of science. *See* Postulation
Mathematics role in modern science, 25–26
"Matthew effect," 166, 181
Matzat, Uwe, 164
McKim, Geoffrey, 35, 36, 166, 167
Measurement of complex observable relations. *See* Experimentation
Meho, Lokman, 169
Merton, Robert, 177, 186, 233n7
Metropolit case, 110
Meyer, Eric T., 97, 149
Microsoft Research, 141
Middleware, 77
Modeling, 24
Monopoly, 181
Multidisciplinarity, 46, 132, 142, 206
Multi-level, multi-domain intelligibility devices, 28
Mutual dependence, 10, 28, 32–33, 92, 120, 154, 168, 205

National Center for Biotechnology Information, 193
National Defense Radio Establishment law, 109
National Endowment for the Humanities (NEH), 147
National Grid Service, 47
National Institutes of Health (NIH), 171, 193
National Science Foundation (NSF), 46, 47, 211
Net neutrality, 234n18
Networked science, 215–216
Newton, Isaac, 34
Ngram Viewer, 37, 154
Nicholas, David, 162–163
NodeXL, 141, 192
Nongovernmental organizations (NGOs), 175, 176, 177, 179, 180
NorduGrid, 137
Novartis Institute for Biological Research, 73, 75

Objectivization, 29
Observatories, virtual, 16–18, 134, 137–138
Ocean Tracking Network, 212
Office of Cyberinfrastructure (OCI), 46–48
Old Bailey Proceedings Online (OBPO), 150, 151–152, 232n6
Online surveys, 46
Ontological engineering, 79
Open-access publishing, 165–166
Open Middleware Infrastructure Institute, 176, 233n2
Open science, 184–186, 233n9
 attention space and, 181–184
 battle between closed and, 175–176
 limited impact of open-research infrastructure and, 181–184

norm of openness and, 177–181
as precondition for modern science, 175
pushes for, 176–177
Open Source Advisory Service, 176
Open-source software, 178
Ordnance Survey, 188–190
Organized skepticism, 177
OSS Watch, 176
Oxford Internet Institute, 3
Oxford University, 85, 147, 150, 152, 232n7

Path dependence, 181
Phenotypic data, 121–122
Physics
  EGEE and high-energy, 135–137
  large-scale data and, 9–10
Pinch, Trevor, 36
Podvinec, Michael, 74, 75, 77
Popper, Karl, 19
Postulation, 23, *24*
Practice-based universality, 28–29
Practices, 214–215
Price, Derek de Solla, 96
Privacy issues, 194–195
Probability, 24
ProQuest Dissertations & Theses, 55, *56*
Proteomics project, 74–75
Psychiatric genetics, 12–13, 101–105, 112–113
Public engagement via citizen science, 10–11
PubMed Central, 2
Pynchon, Thomas, 16–17, 86–88
Pynchon Wiki, 16–18, 86–93, 119–120, 124, 127
  collaboration on, 207–208
  scientific style, 198, *199*, 202, 205

RePlay project, 149–151, 232n5
Research. *See also* e-Research
  disciplinary differences in transformation of, 206–208
  e-research transforming, 8–18
  knowledge versus, 6
  STINs and momentum of digital, 208–211
  users, 36–38
  visibility and access, 18–19
Researchers. *See* e-Researchers
Rosenbaum, Howard, 40
Rowberry, Simon, 91

Salinger, J. D., 88
Sathe, Nila, 161
Savage, Mike, 5, 145
Scalability, 96
Schroeder, Ralph, 6, 105
Schwede, Torsten, 75, 77
Science Commons, 176–177
Science(s)
  aggregating many eyes for, 81–86
  autonomization of, 220
  cumulative, 203–204
  disciplinary differences in e-research and, 128–134
  e-research case studies in, 134–138
  norm of openness and, 177–181
  web, 220
Scientific American, 164
Scientific/intellectual movements (SIM), 41, 211–216, 221
Scientific styles, 23–34, 197–206
  digital components of, 24–25
  high-consensus rapid-discovery science and, 26–28
  mutual dependence and task uncertainty and, 33
  postulation, 23, *24*, 26
  practice-based universality and, 28–29

Scopus, 48–49, 54, 55, *56*, 169, 170, 227n3, 227–228n9
Second Life, 189
Sensors, 118–119
Shinn, Terry, 28–30, 50, 76, 142
Skepticism, organized, 177
Skype, 86
Slaughter, Anne-Marie, 180
Sloan Digital Sky Survey, 2, 10, 82
Slow Learner (Pynchon), 88
Social informatics, 34–38
Social Informatics Data Grid, 47
Social Media Research Foundation, 141
Social network analysis (SNA), 16, 32, 192–193
Social phenomenon, the Internet as, 3
Social sciences
  big data and Google, 143–145
  case studies in, 138–145
  as cumulative, 204
  disciplinary differences in e-research and, 128–134
  e-research publications in, 54
  e-social science, 45–46, 138–145
  VOSON, 16–18, 140–143
Social Sciences and Humanities Research Council, 147
Socio-technical interaction network (STIN) approach, 35–36, 208–211, 217
Soft power, 180
SPLASH. *See* Structure of Populations, Levels of Abundance, and Status of Humpbacks (SPLASH)
Statistical analysis. *See* Probability
Stauffer, Andrew, 154
STIN. *See* Socio-technical interaction network (STIN) approach
Stokes, Donald, 80

Structure of Populations, Levels of Abundance, and Status of Humpbacks (SPLASH), 15, 97–101, 127
  scientific style, 198, *199*
  in shift from small to big science, 112–124
Styles of science. *See* Scientific styles
Supercomputing, 51–52
Sweden, data sharing in, 105–112
Swedish National Data Service, 13–14, 111
  scientific style, *201*
Swedish Research Council, 106–108
SwissBioGrid, 14–15, 72–81, 119
Swiss Institute of Bioinformatics, 74
Swiss National Supercomputing Centre, 73

Tancer, Bill, 144
Task uncertainty, 33
Taxonomy, 24, 119
  interpretive, 198
Text Creation Partnership (TCP), 55
Textual analysis, large-scale, 152–154
Thelwall, Mike, 165, 169
Thomas, Arthur, 74
Transformation of research, disciplinary differences in, 206–208
Transformations of knowledge, variety and homogeneity in, 216–218
Transmission Control Protocol/Internet Protocol (TCP/IP), 5
Trust and privacy, 194–195
Twitter, 86, 145
Typology of Internet research, 3–4

UK National Centre for e-Social Science (NCeSS), 45, 47, 66, 211
Underlying technologies of the Internet, 4
Universalism, 177, 178, 186
Users, 36–38

Index

defining, 125–126
end, 126

Vallas, Steven, 80, 81
Van Arkel, Hanny, 10
Vasileiadou, Eleftheria, 165
Virtual London, 187–190
Virtual Observatory for the Study of Online Networks (VOSON), 16–18, 140–143, 192
  scientific style, *201*, 202
Vision of Britain, A, 151, 232n9
VOSON. *See* Virtual Observatory for the Study of Online Networks (VOSON)

Wallace, David Foster, 17
Waller, Vivienne, 143–145, 153
Walsh, John, 168
Ware, Tim, 89
Web archives, 190–193
Weber, Steven, 178
Web of Knowledge, 37
Web of Science, 169, 170
Webometric approaches, 140
Web science, 7, 220
Websphere, 166
Weisenburger, Steven, 87, 90
Whitley, Richard, 32–33, 119
Wiegland, Gordon, 48
Wikipedia, 3, 19, 86, 89–90, 91, 145, 155
Wilkinson, David, 164
Willinsky, John, 179
Wired Magazine, 196
Wood, Michael, 87
World Internet Project, 33

Yahoo, 162
Yang, Kiduk, 169
YouTube, 149

Zeitlyn, David, 66
Zooniverse, 11, 83–86
Zurich Functional Genomics Center, 73

www.ingramcontent.com/pod-product-compliance
Lightning Source LLC
Chambersburg PA
CBHW021349300426
44114CB00012B/1148